Karl Sabbagh is a writer and documentary maker, living in Oxfordshire. He has written twelve non-fiction books, as well as making documentaries for the BBC, Channel 4 and PBS. He was educated at King's College, Cambridge, where a line in an obituary notice led to the unravelling of the 'rum affair' at the heart of this book.

Books by Karl Sabbagh

The Living Body
Skyscraper
Magic or Medicine (with Rob Buckman)
Twenty-first Century Jet
A Rum Affair
Power into Art
Dr Riemann's Zeros
Palestine, A Personal History
Your Case Is Hopeless
Remembering Our Childhood
The Hair of the Dog
Shooting Star
Britain in Palestine
The Trials of Lady Jane Douglas

A Rum Affair

A True Story of Botanical Fraud

Karl Sabbagh

BIRLINN

This edition first published in 2016 by
Birlinn Limited
West Newington House
10 Newington Road
Edinburgh
EH9 1QS

www.birlinn.co.uk

First published in 1999

ISBN: 978 1 78027 386 0

British Library Cataloguing-in-Publication Data
A catalogue record for this book is available from the British Library

Typeset by Hewer Text UK Ltd, Edinburgh
Printed and bound by Grafica Veneta
www.graficaveneta.com

Contents

	List of Illustrations	vi
	Foreword	vii
	Acknowledgments	xi
1.	'A Curious Episode'	1
2.	Fields of Inquiry	9
3.	The Professor and the Don	25
4.	Rumours	48
5.	The Plan	68
6.	The Forbidden Island	83
7.	'Quoth the Raven . . .'	102
8.	'Is Such a Thing Done?'	122
9.	The Aftermath	145
10.	Seeds of Doubt	156
11.	'A Total Muddlehead'	174
12.	The Matter Subsides	204
13.	'Broken, Lost or Never Collected'	217
	Epilogue: Two Minor Mysteries	229
	Appendix	243
	Notes	247
	Index	253

List of Illustrations

1. Professor John Heslop Harrison.
2. John Raven.
3. Alfred Wilmott.
4. Maybud Campbell.
5. Heslop Harrison on a fieldwork trip.
6. A party of botanical field workers from Newcastle University on a trip to Rum.
7. Heslop Harrison botanising on Rum.
8. Heslop Harrison took botany students every year to Rum.
9. Heslop Harrison's house in Birtley, Tyne and Wear.
10. King's College, Cambridge.
11. Kinloch Castle on Rum.
12. Kinloch Castle, built by the Bullough family.
13. A bronze monkey-eating eagle in the Grand Hall of Kinloch Castle.
14. Plants on the Rum might have provided evidence for Heslop Harrison's theories.
15. The stream where John Raven made several suspicious discoveries.

Foreword

A mysterious kind of parallelism is at work in the story Karl Sabbagh tells so brilliantly in this book. The occasion, as one of his informants says, may have been 'the greatest scandal of twentieth-century botany', but the energy of the drama is provided by the collision of principles and the meeting of two men driven by deeply opposed ways of being.

It is, in a way, a class drama: John Raven, of Trinity and King's College, Cambridge, the son of a dynasty of intellectual distinction on both his parents' sides, stretching back into the eighteenth century, comes up against John Heslop Harrison, a self-made man, of King's College, Newcastle, the son of an ironworker and the first in his family ever to live by the work of the mind.

But that crudely binary social division scarcely embraces the whole of the story. Gamekeeper and poacher are deeply intertwined here. Heslop Harrison was the older, establishment man, one of the most prominent professional botanists of his generation, the founder himself of a dynasty of scientists that is still going strong. Raven was young, an amateur, a great sleuth after rarities but by training a classicist, steeped in the philosophy of Plato and the pre-Socratics, for whom botany was a fascination and obsession but of his spare time. Harrison was a man of distinction and immense self-assertive seriousness; Raven, at least in part, a witty, playful, charming joker. Of deep social principle, he would have been the very last person to embrace the idea that inheritance was somehow the definer of his or anyone else's life.

The intriguing aspect of the moment in which these two lives met and wrestled with each other is that the very ground on which they fought – were Heslop Harrison's discoveries of rare plants on Rum true? – was itself bound up with these questions of self-invention and inheritance. Were those plants, as Harrison claimed, the continuation of long presence and persistence? Or were they, as Raven established, newly invented there? Or, to put it in a more polarized form, were the rare discoveries on Rum more like the heir to the Cambridge intellectual aristocracy or more like the Newcastle ironworker's son? Each embraced his opposite: Harrison championed them as Ravens; Raven revealed them as Harrisons.

Harrison was also, it turns out, a Lamarckian, believing in a pre-Darwinian way that an organism could transform itself during its own life, absorbing influences from the environment, and somehow transmitting those changes to its own offspring. Nature itself could in that way become part of the Harrison conspiracy. You could make it up and once you had made it up, what you had invented became part of how things were. Genetic inheritance was no prison, as the Darwinians insisted. The individual could mould his own destiny. Reality was not there to be discovered; it could be transformed. This foundational intellectual confusion in Harrison was what allowed the great evolutionary geneticist John Maynard Smith to call him not a fake but 'a total muddlehead'.

I have a confession to make: John Raven was – is – my father-in-law. I prefer 'is' because he remains a presence in the life of his children. I am married to his daughter Sarah who loved him and learned her love of flowers from him, botanizing with him as a girl, making a list of species with him that they had seen and identified together in Cambridgeshire and Ireland, the Dolomites and the West Highlands. I never met him, but she and I have gone together with his articles in hand to the hills around Asolo and the White

Mountains in Crete, using his descriptions to find the wild tulips that grow there in patches of dazzling, life-memorable beauty among the limestone crags. It is a connection with nature his father had given him and is, I am sure, one of the most valuable things any parent could give any child.

So I come to this story from one side. John Heslop Harrison's confusion and fakery looks little more than pitiable in the light of John Raven's analysis and destruction of it. The charming, joke-filled, teasingly gentle and amused, ironic man so often described by his friends and family disappears in this story under something cooler and more unforgiving. The idea that this is the story of a patrician, amateur triumph over professional bluster – Raven as a kind of botanical Lord Peter Wimsey – is not really good enough. Max Walters, Raven's co-author on the New Naturalist classic *Mountain Flowers* (1956) described 'the extraordinarily accurate information which John accumulated with characteristic clarity and precision in the famous card index of rare plants.' All British botanists came to recognize, Walters said, that 'the best way to see any rare British plant was to ask John Raven's advice.'

His triumph wasn't amateur luck, nor was it pure cerebration. He had an astonishingly good nose for the site of a wild flower and would dive unerringly for the place a rare plant was growing by understanding above all about habitat, context and the conditions of vegetable life. But in this gift for a supremely uncluttered processing of complex information, there was a connection to his work in ancient philosophy. As all his friends have said, he was never a man for grandstanding, or for adopting big pompous positions on truth and reality. His whole frame of mind was provisionalist, settling for what seemed to be true until shown to be otherwise. This lovely Socratic quality, as if Montaigne had taken up botany, is at the heart of this book and of John Raven's life: a radical scepticism, a relish for the absurd, a nose for the truth and a needle for any wobbling balloon with even a whiff of pretension about it.

In that there is another kind of parallel here: if John Heslop Harrison's tragedy was an addiction to pretentious self-invention, John Raven's was a pervasive scepticism that extended even to his own abilities and worth as a person. His friends have described how, as he grew older, Raven's 'perilous tensions and contradictions' and the 'profound sense of his own inadequacy' led him to addictive smoking and drinking and in 1980 to an early death in his mid-sixties. You could make a connection here: it may be that the sceptical cast of mind that allowed him to see through what Harrison had done on Rum also came to deny him in the end any sense of ease in the world.

Better, though, to remember the laughter which that wonderful unpretentiousness also gave him. The novelist Celia Buckmaster, wife of the anthropologist Sir Edmund Leach, the provost of King's, remembered a moment when she was with the Ravens at home in Ardtornish in the west Highlands. Because his emphysema prevented him from going on long walks, John used to like to send his friends and pupils out to remote glens to gather rarities from spots where he thought they might be found. One day a young man, arriving back in triumph from his expedition, brought John a wild flower for identification. 'Yes,' he said carefully, 'the common dandelion – *but a very good specimen.*'

Adam Nicolson
March 2016

Acknowledgments

I could not have embarked on the writing of this book without the wholehearted support of Faith Raven. I needed her permission first to read John Raven's file in the library of King's College, Cambridge, and she gave it willingly once she discovered that I was a Kingsman. When I said I'd like to write a book about the Rum affair, she provided me with access to other papers, contact names, tape recordings, photographs, and interviews. I am equally grateful to Pat Heslop Harrison, who from my first contact with him put no obstacles in the way of my investigating the allegations against his grandfather. In fact, he displayed a genuine interest in the topic of scientific fraud and gave me permission to quote from any of his grandfather's correspondence and publications. The third important source of help and support was Max Walters, who unwittingly set me on my path by writing a paragraph in John Raven's obituary. He has since been a vital source of botanical advice and has shared with me the task of trying to solve some of the puzzles presented by this odd story as I uncovered it.

I have benefited from conversations and correspondence, sometimes on several occasions, with David Allen, Gill Bray, Jasper Brener, Mary Briggs, Tim Clutton-Brock, Alan Davison, Hal Dixon, Garth Foster, David Harrison, Edward Katkin, John Maddox, Harvey Marcovitch, Peter Marren, Heather Marshall, John Maynard Smith, John Morton, Richard Pankhurst, Chris Preston, Miriam Rothschild,

Brian Silman, Colin Welch, and Peter Wormell, none of whom, unless clearly identified in the text, should be associated with any specific observation or comment.

I am grateful for the assistance of Peter Jones and Jackie Cox in the King's College library, and John Thackray in the archives of the British Museum (Natural History). I am also grateful to Professor Stephen Bann and Dr Stephen Blackmore, out of whose chance meeting came further embellishments to the strange story I tell in this book. And even while the book was being prepared for the printer, Hugh Raven, John Raven's son, suddenly uncovered – and kindly lent me – a battered folder which provided one or two last-minute insights into the story.

Andrew Rosenheim of the Penguin Press took time off from his busy life as a publisher to edit the original manuscript with me and made many helpful suggestions as to structure and style.

I am grateful to the following, who gave permission for passages to be quoted in the book or photographs to be reproduced:

American Association for the Advancement of Science
University of Newcastle upon Tyne
The Royal Society
Nature
The Natural History Museum
Macmillan, UK

I

'A Curious Episode'

Every year, King's College, Cambridge, sends out its annual report to graduates of the college. In an average year, the information it provides is usually of little interest to anyone on the outside, and hardly more riveting to the people it is produced for: how well the college did academically (usually very well) and sportingly (usually derisory); which graduates gave copies of their new books to the college library (Jan Pieńkowski has donated another *Meg and Mog* book, for example); who landed a prestigious job (Her Majesty's Ambassador to Venezuela, Artistic Administrator of the Detroit Symphony Orchestra); news of college servants (Mrs Hoye, bed-maker, retired after twenty-eight years' service); and so on. Most of the report is taken up with obituaries of fellows and graduates, which were written for many years by Patrick Wilkinson, a Fellow of the college, who clearly treated the task as the opportunity to create a minor art form.

Although I am a voracious reader, I do not make a practice of reading obituaries. But there was a unique quality about the obituaries of Kingsmen that meant I usually tried to find time to sit down and read them within a day or two of the arrival of the annual report. Not to put too fine a point on it, they contained some good examples of donnish wit. For example:

'In College, he organised nocturnal races, naked except for gowns, on the older Fellows' tricycles.'

'His most vivid recollection of the College was to be of the Provost's wife.'

'Baron Corvo . . . braved travelling in his sidecar despite his insistence (before the days of traffic lights) that the faster you drove over the crossroads the smaller was the mathematical probability that you could have a collision.'

'After passing second MB he served as a surgeon sub-lieutenant in destroyers, based on Malta, and remembered his ship shelling the British lines in Gallipoli, a mistake he attributed to his commanding officer's breakfasting on port and pineapple chunks.'

'Meanwhile, he made heroic efforts of cycling to spend as much of his leave as possible with the choirboys of Canterbury Cathedral.'

'He developed blackwater fever and qualified for the *Guinness Book of Records* by achieving the highest known nonfatal temperature.'

'He radiated fun and he never lost the sense of mischief which had prompted him as a small boy to write "bugger" with a pin on a burgeoning vegetable marrow in a vicarage garden.'

'Due to an inspired misprint in *The Times* obituary he was described as being "survived by many nephews and pieces".'

You probably get the idea.

In 1980, the usual brown envelope dropped on my doormat, and a few days later I sat down to browse the obituaries. One of the dons who had died that year was a classics tutor named John Raven. He had been senior tutor when I was an undergraduate. Truly the only thing I remembered about Raven was that, at a welcome sherry party for new students, he had demonstrated an ability to move from a standing to a cross-legged position in one smooth movement while keeping his torso entirely vertical.

Raven's obituary had its share of wit, but one particular paragraph intrigued me:

In 1954 he was at the centre of a curious episode. A reputable biologist had recorded finding, mainly on the Isle of Rhum,* several plants not previously found in Britain. The botanical world was surprised, not to say suspicious. John went to investigate. His report was deposited in Trinity College Library and has never been published. John was indignant with editorial scientists who thought that 'not having unpleasantness within botanical circles' was more important than truth; but enough was made public to secure that the plants were quietly dropped from later editions of British flora. Kingsmen recalled the exposure of T.J. Wise by J.W. Carter of the year 1924.

For seventeen years this elliptical story stuck in my mind as worth exploring further, but I did nothing about it. Then, in 1997, when I had a prominent botanist to lunch – the first such occasion in my life – I mentioned the story, and the botanist told me, in a rather unforthcoming way, that the allegations of fraud investigated by Raven concerned a Professor John Heslop Harrison of Newcastle University. Immediately, my interest intensified. Giving the alleged culprit a name put flesh on his bones and made me want to know more about him.

I should say straightaway that it was not easy to discover what that name actually was. The professor had started as a simple Harrison, with Heslop as a middle name, and at some point decided to adopt the Heslop as part of his surname. But there were times when some people called him one name and others called him the other simultaneously.

My botanical friend's slight reluctance to expand on the story also increased my interest. He was concerned that Heslop Harrison had a son who was still alive and had himself become a distinguished botanist. He gave the impression that the whole profession

---

* 'Rhum' was the spelling used up to the 1970s. See footnote on page 68.

knew about the accusations but that, on the whole, this was never openly acknowledged.

We are none of us responsible for the sins of our fathers, and yet apparently, after *this* father's death an entire generation of botanists had tiptoed around the accusation to avoid offending his son, who, I presumed, had played no part in the misdeeds, if misdeeds there had been.

So, where to start? Patrick Wilkinson was dead by now, but, as the obituary said, the 'report was deposited in Trinity College Library . . .'

Oh no, it wasn't. A librarian at Trinity acted on my phone call, spent a couple of days investigating, then called back to say that there was no trace of the report. 'Have you tried King's? It may be there,' she said. I hadn't, I did, and it was.

There was a certain amount of caginess on the part of the King's librarian. 'You'll have to get Mrs Raven's permission,' he said. Mrs Raven – Faith Raven, John Raven's widow – turned out to be a woman who eschewed small talk and spoke her mind. Having established that I was a Kingsman, and therefore presumably 'a safe pair of hands,' she agreed that I could read her husband's report.

In a light, airy room on the second floor of the library, several people sat almost motionless at tables, turning the pages of rare books or carefully perusing sheets of manuscript. The quiet scratch of a pencil – no pens allowed – would occasionally enliven the silence. Researchers came here from all over the world to consult the papers of E.M. Forster or John Maynard Keynes, or any of the hundreds of King's fellows or graduates who have left material to the library. Already waiting on 'my' table was a faded beige manila folder with 'not to be looked at in the life of J. Heslop-Harrison (Jnr)' written on the outside. This interdict introduced a cloak-and-dagger element into the business since, after all, I *was* looking at it 'in the life of J. Heslop-Harrison (Jnr)', the son whom the nation's botanists, it seemed, wanted to avoid offending.

The first things I came across in the folder were photocopies of handwritten letters, along with carbons of typewritten letters. The carbons were from Raven, and the handwritten copies were replies from Heslop Harrison. There were also miscellaneous letters from Raven to others, including one, dated 1960, to a friend. It had accompanied a copy of the report and said, 'I'd really like it kept under lock and key until both its hero and I are safely in our graves.' (Calling Heslop Harrison the hero of the story, when he was anything but, turned out to be a typical example of Raven's gentle irony.)

The bulk of the folder was taken up by a foolscap envelope. In the envelope was John Raven's report, in his small, neat handwriting. It began: 'This report will inevitably be of considerable length, and much of it will, I'm afraid, be rather dull . . .'

When I finished reading, twelve thousand words later, I could not agree with Raven's initial warning. It was not dull at all. It was a cool, reasoned, humorous, and explosive indictment of chicanery of a high order against a man who, throughout most of his career, was believed by non-botanists to be a model of academic rectitude, rigorous research methodology, and distinguished discoveries.

At the heart of the report was an accusation, made quite explicitly, that Professor John Heslop Harrison of Newcastle University had, at some time in the 1940s, transported alien plants to the Isle of Rum in the Inner Hebrides and planted them in the soil. He had then, Raven alleged, 'discovered' the plants and claimed that they were indigenous to the area and that he was the first to come across them. The report dealt with one specific group of plants that had come under suspicion, but the implication was that a whole series of other plants, described by Heslop Harrison and his students in the Inner and the Outer Hebrides over the previous decade or so, were also 'planted' and therefore could not be considered 'discoveries,' as had been reported in academic papers by the professor and his students.

The Raven report presented a long list of Latin names of the plants in question – *Epilobium lactiflorum, Erigeron uniflorus, Carex capitata, Carex bicolor, Lychnis alpina* – and as I read it, images flashed before my eyes of tiny, brightly coloured edelweiss, ornate, exotic orchids, primroses, pansies, and buttercups. It was some time before I realised that the plants at issue, which formed the focus of so much passion, were grasses, sedges, or rushes, mainly – to a non-botanist unremarkable. One in particular, the sedge *Carex bicolor*, was both the most unusual of Heslop Harrison's discoveries and one of the least flamboyant.

Over the next few weeks, I started to talk and correspond with a number of people who might have been expected to know of the events in the Raven report and the broader issue of fraud by Heslop Harrison. The botanists I began to consult were divided into three groups: those who said they knew little or nothing of what I was talking about; those who believed that Heslop Harrison had done what Raven said he did and was lucky to have got away with it for so long; and those who believed that he did *something* a little under-handed or incompetent but that because he was such a good researcher most of the time, kind to his students, and an autodidact who deserved admiration for what he had achieved, his reputation should be left intact. None of those I spoke to, apart from close friends of Raven's, had seen the Raven report or knew much about what was in it.

It was also difficult to find anything in print about the contro-versy. In the file was a reprint of a letter from John Raven to *Nature*, the most authoritative British science journal, published in the issue of 15 January 1949, eighteen months after a trip Raven had made to Rum. How and why Raven had written this wasn't clear to me at the time. But to anyone not in the know, it was far from providing evidence of a smoking gun, let alone of the hand that had fired it. The only reference to Heslop Harrison was where Raven said '. . . thanks to the kindness of Professor J.W. Heslop Harrison . . .

I was enabled to see some at least of [Rum's] most interesting plants.' To someone who had read Raven's report, this would seem on a par with thanking the embezzler for allowing you to see the forged cheque.

John William Heslop Harrison was Professor of Botany at Newcastle University at the time Raven wrote his report. He was then aged sixty-seven, and plants and insects had been a lifelong passion. His mother had 'the green fingers of a born gardener' and apparently was the only person to whom the young Heslop Harrison would entrust his moths when he was away from home.[1] His first job was as a secondary-school teacher, and for twelve years all of his natural history work, leading to scientific papers in the specialist journals, was done outside his working hours. It was that research, and the commitment that underlay it, that led to his first university post, as a researcher at the University of Durham.

A phrase that was often used later, much later, by people who spoke to me about Heslop Harrison was that he was 'a miner's son.' Not quite accurate, but it has resonances that go far deeper than a bald description of his father's education. Such an expression implies the progression that underlies the American phrase 'from log cabin to the White House,' and I have the impression that it might be something Heslop Harrison said, or at least thought, of himself: 'from miner's son to university professor.'

There seemed to be three interwoven worlds that formed the background for the events of the story. There was, of course, the scientific world. This is a world in which hypotheses are formed, data gathered, and theories revised or consolidated, all in an environment from which human emotions, motives, and frailties are meant to be excluded. Whatever had gone on, whatever Heslop Harrison had done, was confined to one part of this scientific world: what is now called biogeography, which deals with how and why particular plants are found in particular places. Heslop Harrison and Raven played subtly different roles on this stage because, for all his botanical expertise, Raven, unlike Heslop Harrison, was not a professional scientist. His 'day job' was as a classics scholar and teacher, an authority on the Greek philosophers. All of his botanical activities were pursued as an amateur, but they were of a high order.

Then there was the academic world, split into the two spheres that Heslop Harrison and Raven inhabited: 'Oxbridge' – Oxford and Cambridge – and what were known as the 'redbrick' universities, in towns and cities throughout the rest of Britain. Oxford and Cambridge have always attracted accusations of elitism. And John Earle Raven's family background – son of the Master of a Cambridge college, with academics, schoolmasters, and clerics in his background – meant that he would naturally go to a Cambridge college. His undergraduate work at Trinity led effortlessly to a first-class degree, followed by a fellowship in classics at the same college, awarded in 1947. Heslop Harrison, on the other hand, was the son of an ironmaker in a small village in the north of England, and his education involved financial sacrifices by a family that wasn't particularly well off. For him to contemplate a university education at all was quite a step. Emerging redbrick colleges, such as King's College, Newcastle – initially part of the University of Durham and later the nucleus of the University of Newcastle – lacked the history, tradition, and even attractiveness of the older universities yet felt themselves just as capable of competing with the best in the land.

They were very different men socially, Raven and Heslop Harrison, but they were linked by the third world, the world of natural history, and specifically of plants. Although there was an overlap with the scientific world, it soon became clear that many botanists, like Raven, were not necessarily scientists, or not particularly interested in science, but *were* passionate about plants. I felt it was a passion I would have to learn to understand, whether or not I could share it, if I was to get much deeper into the story of the Rum affair.

2

Fields of Inquiry

Botany does not have the image of a serious science. Physicists, chemists, biologists, even mathematicians, represent in the public mind the potential to do great good – or great evil. But you don't expect botanists to win Nobel Prizes and, just as important, you don't expect them to destroy the world one day. Some of these reduced expectations derive from the blurred border between natural history and botany. Anyone can do natural history, and many of us have in our childhood – collecting minnows in jars, pressing wildflowers in albums, gathering mushrooms and toadstools, and generally taking a delight in observation, classification, and discovery. Some young natural historians become older ones and move professionally into the biological sciences; others, like John Raven, never stop being amateurs but make a major contribution to the professional field.

But what does 'major contribution' mean to botany? Again the question arises: How serious a science is it? Supposing we didn't know the precise distribution of British sedges. Would it really matter? Would it in fact matter whether *Carex bicolor* and other plants 'discovered' by Heslop Harrison were natives of Rum or not? Why did I find it amusing when one of the botanists I spoke to early on said in passing that 'there was some kind of controversy about pondweeds'? But, of course, every single other scientist has preoccupations that, while arousing his or her own passions, produce boredom or incomprehension in everybody else.

I have read a lot of scientific literature, although to call it literature in the usual sense of the word is to give it more credit than it deserves. Most scientific papers serve very specific purposes, and providing enjoyment to the reader is usually not one of them. Like all scientific fields, the level of detailed knowledge required to be a good botanist seems formidable to the non-specialist. But it has to become second nature to the fieldworker, who can't get past first base without the ability to see and name fine distinctions of plant morphology and to place individual plants in the overall scheme of botanical classification. And yet the botanist is not in a laboratory, where technical terms and experimental procedures make up the entire discourse, but in the real world, where science and life combine. This combination is reflected in the way in which the specialist terminology of the science merges effortlessly with the reality of the messy world of mountains and earth, bogs and weather.

Two quotes convey these merged worlds. From Raven:

> . . . the Storr and the Quirang . . . are among the most friendly [of British mountains]. Every cliff, rock and scree on the latter half of this short climb has a rich and varied mountain flora. Northern Rock-cress, moss campion, mossy cyphel, the saxifrages *S. oppositifolia* and *S. hypnoides*, rose-root, the grass *Poa glauca*, and green spleen-wort (*Asplenium viride*) are all abundant . . . There can be very few hills in Britain of so lowly a height that can boast so rich a mountain flora.[2]

And Heslop Harrison:

> *Carex bicolor* All. Fl. Ped
> The root stock is short, creeping and somewhat stoloniferous, one in our possession having no fewer than eleven branches . . .

The upper three spikelets are shortly stalked and approximated, whilst the fourth (or lowest!), with a very long peduncle, is distant. The upper bracts have rounded bases and short green pointed tips; the low bract is sheathing, long and leaflike. The glumes are blackish, with a green keel; the utricles are rounded at the summit, without beak, whitish, erect, obovoid, slightly rough, with the nervures scarcely visible, whilst the achene is obovoid, plano-convex and punctate.[3]

Faith Raven told me that she delayed marrying John Raven for years because she was so bored by botany. I found this very surprising at the time. She has one of the most interesting gardens in the county and throws it open to the paying public at weekends and holidays, and I somehow equated gardening with botany. But it became clear on reading some of John Raven's writings that an interest in botany is very different from a love of cultivating plants. In fact, there seems sometimes to be very little overlap.

Good botanists are interested to the point of obsession in tiny details of plant anatomy and physiology; they will sacrifice time, money, and comfort to seek out a plant they have never seen before, or a familiar plant in an unfamiliar location; and they see beauty, wonder, and delight in a plant that you or I – if you're also a non-botanist – would crush underfoot without realising we'd just damaged a rare specimen of *Hieracium praetenerum*.

One of John Raven's friends, Tim Clutton-Brock, who has spent over twenty years of his professional life studying the red deer of Rum, had an interesting theory about how botanists differ from other scientists. He contrasted them with zoologists: 'My impression is that botanists are really fascinated with the details and the facts, whereas zoologists like working in bigger, broader brush, theoretical frameworks within which they structure how they see the world. Botanists tend – or have tended in the past – to see the world with a more microcosmic view, more closely related to the facts.' He saw this distinction, if it exists, as one possible difference between Raven and Heslop Harrison.

Heslop Harrison's first university post was actually as lecturer in zoology, and his botanical interests were always interwoven with research on insects. As we will see, it is likely that the motivation for some of his key research was, indeed, to promote and justify grand biological theories.

But how true is it that 'botanists are fascinated with the details'? David Allen, a man who has devoted his life to the history of British botany, describes one of the British botanical greats, H.C. Watson, as having an 'accountant-like approach to the study of why plants grow where they do.'[4] And Ernest Rutherford, the New Zealand physicist, was scornful of some branches of science, including botany, which he described as stamp collecting.

Raven's interest in botany was very much in the English tradition of the amateur who contributes as much to a field of knowledge as the professionals. Ornithology, geology, palaeontology, and even astronomy have all benefited from the passions of people whose regular occupations are entirely different and sometimes more mundane. Raven's botany was also in the tradition of his father's interest, and he, in turn, was an admirer of the man who was the founder of the systematic study and classification of plants in Britain. Canon Charles Raven was Regius Professor of Divinity at Cambridge and became Master of Christ's College. In 1942, he wrote a biography of John Ray, a seventeenth-century naturalist. It was a work that could have been written only by someone who had a fluent knowledge of Latin and a detailed familiarity with the natural history of Britain. Charles Raven had both, and in spite of all the other demands on his time during the early years of the Second World War, he managed to find time to write a book of five hundred closely printed pages, with as many as a dozen footnotes per page, showing how important Ray's ideas were for the sciences of botany and entomology.

Charles Raven wrote in 1942: 'If to-day it amazes our continental neighbours to find the greatest English newspaper reporting the annual arrival of Cuckoos and Swallows or the capture of a Camberwell Beauty and the spread of the Comma, the development of that trait in our national life is due to John Ray far more than to any other man.'[5]

In his translation of a Latin paragraph from the preface to Ray's first book, published in 1660, Charles Raven describes a passion that might as well be his own, or his son's:

> I had been ill, physically and mentally, and had to rest from more serious study and ride or walk. There was leisure to contemplate by the way what lay constantly before the eyes and were so often trodden thoughtlessly under foot, the various beauty of plants, the cunning craftsmanship of nature. First the rich array of spring-time meadows, then the shape, colour and structure of particular plants fascinated and absorbed me: interest in botany became a passion.[6]

A couple of pages later Ray continued:

> Surely we can admit that, even if, as things are, such studies do not greatly conduce to wealth or human favour, there is for a free man no occupation more worthy and delightful than to contemplate the beauteous works of nature and honour the infinite wisdom and goodness of God . . . Of course there are people entirely indifferent to the sight of flowers or of meadows in spring, or, if not indifferent, at least preoccupied elsewhere. They devote themselves to ball-games, to drinking, gambling, money-making, popularity-hunting.[7]

(It is a wonderful confirmation of the stability of human psychology that what struck Ray over three hundred years ago as the major preoccupations of people still dominate social life today. The only omission is sex, which Ray probably considered best confined to the stamens and carpels of the plants themselves.)

It was while contemplating the 'beauteous works of nature' that John Ray noticed a fundamental characteristic of plants that had escaped generations of his predecessors. In a paper contributed to the embryo Royal Society, he reported, 'The greatest number of plants spring out of the earth with two leaves, for the most part of a different figure from the succeeding leaves . . . the seed-leaves are nothing else but the two lobes of the seed.' He then went on to write of a smaller group of plants, those whose 'seeds spring out of the earth with leaves like the succeeding . . . nor have their pulp divided into lobes.'

In the church at Grantchester, near Cambridge, a mosaic floor surrounds the altar. Regularly spaced over it are alternating flower designs. One is a six-petalled lily, the other a five-petalled rose. You might easily pass it by without a glance – just two flowers, chosen by the artist for no apparent reason. But Max Walters, a professional botanist, friend of John Raven's, King's fellow, and co-author of *Mountain Flowers*, mentioned it to me during an early conversation, pointing out that the two flowers symbolised the way in which all flowering plants were divided into Ray's two categories.

The lily and the rose are emblems of Corpus Christi College, Cambridge, patron of the Grantchester church. But coincidentally, these two flowers and the number of their petals represent one of the most fundamental divisions of flowering plants, a division that is crucial for plant classification. The two different types of seedling that Ray noticed are called monocotyledon, for the seeds that produce a single seedling leaf, and dicotyledon, for paired seedling leaves. Nowadays, among working botanists discussing the plant world, the names 'monocot' and 'dicot' trip off the tongue. The 'cot' is short for cotyledon, the seedling leaf. 'If you're a gardener,' said Max Walters, 'you know that an onion produces a single long leaf and the brassica – cabbage – produces a *pair* of seed leaves, so a germinating onion seed is a monocotyledon, or monocot, and a germinating cabbage seed is a dicot.'

In a way, it's not surprising that this fundamental division escaped detection for so long. There are about 250 families of dicots and fifty or so monocots, and if you look at the adult plants that make up each group, they seem to have little in common. Monocots include lilies, rushes, sedges, grasses, irises, orchids, and palm trees. Dicots include honeysuckle, sunflowers, buttercups, roses, mustards, mallows, primroses, phloxes, snapdragons, mints, and geraniums. The characteristics that separate the two groups are the sort of thing you notice only if you get down on your knees and peer at the different components of the plant, preferably with a magnifying glass.

Roots, leaves, flower parts, even the layout of tiny tubes that make up the plant's vascular system – all are linked to this earliest visible sign of difference: one leaf or two at the seedling stage. And there's one other easily visible but long unnoticed characteristic that separates the two groups: the flowers of monocots have three or six petals; dicots have four or five. So the lily in the Grantchester mosaic, with its six petals, is a monocot, and the rose is a dicot. Telling a monocot from a dicot is the nursery slopes of botany; it is the starting point for developing the observational skills that both Heslop Harrison and Raven used to explore the world of flowering plants.

It would be a fascinating diversion to explore why plants have flowers at all. In earlier, pre-Darwin centuries, of course, the purpose of flowers was to enhance the beauty of the world and make it more pleasant for the acme of divine creation – ourselves – by contributing to a colourful and scented environment. The advent of the theory of evolution by natural selection means that we have to look for a more hard-headed answer, one expressed in terms of the value to the species of putting a lot of investment into surrounding the inconspicuous reproductive organs with complex, ornate, and highly visible appendages. And the answer is that the plants that have colourful and imaginatively sculpted flowers are those that depend for survival on attracting insects to them to carry pollen – the male seed – from the interior of the flowers to the eggs of another member of the species, or even to other parts of themselves, to fertilise the flowers and produce the seeds for the next generation.

Max Walters made one other remark that had a similar effect on me as his reference to the lily and the rose. It illuminated a corner of the field of botany in a way that cut through the Latin terminology, the complex naming of parts, and the rarefied arguments about the exact structure of stamens and pistils. 'From the point of view of *Homo sapiens*,' Walters said, 'the wind-pollinated monocots are the things by which we live.'

Now, I have to say I had never appreciated that the beer I drink and the bread, porridge, and rice I eat are all derived from 'wind-pollinated monocots.' But it explains why the grasses, which all cereals are, have undramatic and sometimes insignificant flowers. They have no need to be all dressed up since there's nothing they need to attract. They don't need insects, because the wind carries their pollen from the stamen (the male organ) of one plant to the carpel (the female organ) of another. And with some species the wind doesn't even have to blow the pollen very far.

Max Walters's remark was a reminder that, however much of a hobby botany is – and it was clearly an all-absorbing one for the Ravens – its ramifications penetrate the heart of everyday life, including, directly or indirectly, all the food we eat. The observation and comparison of subtle differences between plants, and the compilation of information about where and how they grow, have a much greater significance than mere stamp collecting.

But in spite of the discoveries about plants that have put botany on a firm scientific foundation, it was becoming clear to me that the central issue in this story was not really a botanical one, in the technical sense. If it had been, then I should have been less interested. Raven's list of plants set out above, in his account of the Storr and the Quirang, was both evocative and meaningless to me: evocative because 'rose-root,' 'Northern Rock-cress,' and 'green spleenwort' all convey homely images derived from our ancestors' everyday vocabulary and reflect rural life in a way that their Latin names – such as *Asplenium viride* for 'green spleen-wort' – don't; meaningless because, I have to confess, I wouldn't recognise any of them on a country walk, or even hazard a guess as to which was which. Then, in Heslop Harrison's quote, the mysterious and really rather intriguing terminology was opaque to me, although I rather enjoyed the Lewis Carrol-esque account of peduncles and glumes, bracts and nervures.

While the exact identification of specific plants was a key part of the story, it wasn't the point of it. Many of Heslop Harrison's most important discoveries were significant not because of *what* they were but because of *where* he said he had found them. The surprising locations of many of his discoveries fitted a pattern that helped substantiate a theory he strongly – some said obstinately – supported.

The reason given by Heslop Harrison for first going to the Hebrides in the mid-1930s was to satisfy a long-held ambition. 'The idea of exploring the remoter parts of Scotland for the purpose of studying their natural history has always had a great attraction for me,' he wrote. The opportunity to do so came with a visit to Raasay, one of the Hebridean isles, a visit he found so stimulating that he decided to organise a complete biological survey of the Inner and Outer Hebrides.

The initial intention of the botanists was merely to record the distribution of plants and animals in the area. But, as Heslop Harrison later described it, when he and his group began to study the plants of the Hebrides, it became apparent 'almost simultaneously with the inauguration of our researches' that there were discrepancies between the plants and animals of the Western Isles and those of mainland Scotland.

'So striking were [the discrepancies] in some respects,' said Heslop Harrison, 'that we were forced to the opinion that certain sections of the two populations differed in origin and history.' The significance of that difference lies in a theory about the origin of Hebridean biology that he developed during his research.

In 1951, Heslop Harrison wrote an article for the *New Naturalist Journal*.[8] It was called 'The Passing of the Ice Age and Its Effect upon the Plant and Animal Life of the Scottish Western Isles', and addressed the question of whether some plants and animals that are found today in the British Isles had survived there from before the last ice age, a period when vast swaths of the planet were covered with ice.

As I read Heslop Harrison's article for the first time, I was immediately struck by a particular characteristic of the language. Where Raven's writing was understated and sometimes witty, Heslop Harrison's was often forceful and opinionated. It reminded me of a man on a soapbox, someone who was giving me no opportunity to think for myself but pressing his opinions on me with an unnecessary degree of verbal violence. The sentence I quoted above, which is from that article, is one of many examples. 'We were *forced* to the opinion,' he says (my italics). Already, any doubter among the readers had been put firmly in his place. Over the next few paragraphs, similar examples occur: 'demonstrated conclusively', 'so striking', 'this we had demonstrated was almost certainly the explanation', 'nothing is more certain', 'explicable only', and so on, and so on.

Scientists are often accused of being know-alls, of making dogmatic statements about truth that are later proved wrong. The most familiar example is the dismissal in the nineteenth century of the idea that meteorites were rocks from outer space that fell from the sky. But in fact nowadays, any good scientist, however much he or she believes that a hypothesis is correct, will couch it in probability terms: 'Thus it is likely that . . .', 'All the evidence suggests . . .', 'There is a high probability that . . .' Even when writing for an audience of non-scientists, very few reputable scientists will say, 'I have proved this to be true . . .' Of course, that doesn't mean that the message won't be received with a greater degree of certainty than it is meant to carry. If a cosmologist says that all the evidence suggests that the universe was created from a Big Bang fifteen billion years ago, and later evidence contradicts that conclusion, he will still be remembered as a scientist who told the world something that wasn't really true.

Even making allowances for the intended readership and for the very general nature of the *New Naturalist* article, Heslop Harrison's piece is a polemic, arguing for a particular theory that he advocated and organising the material to leave no doubts in the reader's mind that the theory was 'the truth'. When a scientist espouses a theory with such passion, you can look at it in two ways: either the discoveries drove him to the theory, or the theory drove him to the discoveries. In the years following the start of Heslop Harrison's researches in the Hebrides, many botanists believed the latter was the case.

So what was the theory that Heslop Harrison was so eager to prove?

Heslop Harrison believed – or came to believe – that the distribution of plants and insects throughout the Hebrides proved that some forms of life had survived during the 1.7 million years that Scotland and the north of England were covered with ice. No plants can live under a blanket of ice, so to establish his theory Heslop Harrison needed two types of evidence: he needed to show that certain areas of the Hebrides were ice-free for the whole period of the Ice Age and that the plants in those areas were truly unique specimens rather than plants that could have arrived on the islands from other parts of the area after the end of the Ice Age.

It is a sign of Heslop Harrison's talents as an all-round natural historian that he was able to delve into his stores of knowledge of plants, insects, rocks, pollen, peat, and mammals, as well as more general ideas of geography and evolution, to draw a picture of the biology of the islands during and after the last ice age to cover Northern Europe.

Over millennia, the world has experienced a series of ice ages, periods when large areas of the surface were covered with a sheet of ice thousands of metres deep. The most widely accepted theory for the cause of the ice ages is based on the fact that there have been changes over millions of years in the tilt of the earth's axis and in the circularity of its orbit. Those changes have been cyclical and led to long periods when some parts of the earth's surface received less light and heat from the sun.

Scientists recognise five major ice ages, the first about 2 billion years ago, then three more about 600 million, 400 million, and 300 million years ago, and the last beginning 1.7 million years ago and finishing only about ten thousand years ago. It was the last, the Quaternary Ice Age, with which Heslop Harrison was concerned, a period when the diversity of life forms on land not covered by ice was greater than it had been during the previous ice ages, and during which there were major effects on plants and animals as the temperature dropped and ice sheets and glaciers formed from water that had evaporated from the sea.

As the cold took its grip, not overnight but over thousands of years, the relationship between land and sea changed, caused partly by climate, partly by the weight of ice pressing down on the land. The process that froze water vapour into ice and deposited it on the land effectively depleted the sea of water that would otherwise condense into rain, run into the rivers, and flow back to the sea. As more water froze and water continued to evaporate from the ocean, the sea level dropped, revealing areas that had been covered by water and expanding the total land area. For an area like the

Hebrides, a series of separate islands before the Ice Age, this process would have turned it into much larger areas of land, possibly even into one interconnecting landmass.

The ice also caused the dry land to increase at the expense of the sea because the areas of land under the main burden of ice were pushed downward, which squeezed some of the underlying rock out toward the areas under less pressure. If we imagine the Scottish mainland as bearing the bulk of the ice, the displaced rock from the centre would have lifted the outer edges, including the Hebrides. There were short periods during the Ice Age, 'interglacials,' when the temperature rose, the ice retreated or even disappeared in some areas, and the coastal contours changed, with the sea level rising, new islands forming, and existing populations being cut off.

The effect on the biology of the area over time would be to isolate then reunite parcels of land, so that there were periods of isolated evolution within a small group of plants or animals, followed by periods when they were in contact with a much wider range of species and variants, with opportunities for interbreeding, and with a much wider range of selection pressures.

One event that Heslop Harrison analysed was the last interglacial period, before the ice came down for the final time. He saw it as having been ushered in by a rise in sea level, as the ice melted, leading to a contraction of the land area of the Hebrides. As the ice on the mainland melted and released the pressure on the underlying rock, the peripheral areas would have sunk in a kind of seesaw action between the underlying rock at the edge and that under the ice. During the interglacial period the milder weather would have led to an inrush of plants and animals from further south, now able to survive in these northern areas. Then came the final period of glaciation. Once again, the sea level fell, the air got colder, ice sheets and glaciers built up on the land, and all plant life in those areas died out.

So far, this is a simplified version of what most scientists believe. But Heslop Harrison built a further element into the story, one that, in his view, led to a unique type of flora in the fringe areas of the British Isles. There were areas, he said, that escaped the full onslaught of the ice, areas where life did not die out but survived in a climate that was colder but not lethal. And those areas included the Hebrides.

Heslop Harrison was convinced that parts of the Hebrides had been practically ice-free during the last ice age. By observing the geology of the islands he could see that certain higher-altitude areas showed no signs of the effect of glaciers, the ice sheets that, as they expanded and contracted and moved under the effects of gravity, could carve grooves, grind up rocks into sand, and generally leave their mark on the underlying land. Such ice-free areas are called nunataks by geologists, and sometimes the Hebridean nunataks happened to have rather unusual collections of plants that, in Heslop Harrison's view, were likely therefore to have originated before the Ice Age and survived throughout the period.

As Heslop Harrison put it, in a stirring roll call:

Thus on scattered nunataks, and on cliff ledges, it is pictured that most of the Alpine plants, exemplified by the sedge *Carex capitata, C. bicolor,* and *C. glacialis,* the wood-rush *Luzula spicata,* the grass *Poa alpina,* the Arctic Scurvy Grass (*Cochlearia arctica*), the Fleabane (*Erigeron uniflorus*), the Alpine Saw-wort (*Saussurea alpina*) and the Norwegian sandwort (*Arenaria norvegica*) amongst the flowering plants, and *Andreaea blytii, A. hartmani* and *Ditrichum vaginans* amongst the mosses, survived the rigours of the Ice Age.

Apart from pre-empting criticism by outbursts of certainty, Heslop Harrison gave no alternative explanation for the distribution of the Hebridean plants, and the unqualified reader might be forgiven for thinking that the argument is cut and dried. But for many botanists, the evidence for periglacial survival, as it is called, is not very convincing, and they believe that an alternative explanation is more likely. What probably happened, they say, is that during the Ice Age, in the warmer areas south of the ice sheet, many different plant types existed that are no longer there today. In the ten thousand to fifteen thousand years since the Ice Age ended, two things might have happened: first, the plants may have spread from southern to northern areas as the ice retreated and the climate warmed; second, this broader distribution of a species might have suffered patches of extinction, sometimes quite extensive, producing the illusion of entirely unconnected patches of flora of the same species.

When Heslop Harrison found plants in the Hebrides that had only ever been seen many hundreds of miles away, he interpreted this as evidence of survival from before the Ice Age.

And he had no doubts. In fact, the awkward phrase 'it is pictured', in the last quote, is uncharacteristically reticent compared with the rest of the article. After citing another list of plants and animals that he and his team had identified as having an unusual distribution, Heslop Harrison writes: '. . . it seems incredible that anyone should appeal to postglacial and recent dispersal of an accidental nature to account for their present British distributions.' Or, in other words, 'it seems incredible that anyone would disbelieve me.' With such strong words does Heslop Harrison pre-emptively dismiss critics in print.

Reading such forceful arguments inevitably leads me to think that Heslop Harrison must have felt a strong need to convince his colleagues. And the fact was that many people didn't agree with him. Max Walters and a colleague published a long paper in 1954[9] arguing carefully, and not forcefully, for the alternative explanation of how pockets of unusual plants came to be found in outlying areas of the British Isles. In the end, in science, it is evidence and not argument that wins over the doubters. And it's important to realise that one reason this argument is more important than other controversies in botany is that botanical evidence is one part of a larger picture and therefore supports – or contradicts – evidence from geology, zoology and meteorology. So Heslop Harrison was not just taking on members of his own discipline when he published his evidence for periglacial survival. He was waving the flag for botany on a broader scientific stage, and it was crucial that the case he made was convincing.

But if the theory is so important and the evidence is difficult to find, it might occur to a scientist that there would be little harm in *manufacturing* evidence – illegitimate though such an action would be – since he, the scientist, 'knows' that the evidence is there and it is only a matter of time before he will discover it for real.

There is no doubt that the discovery of *Carex bicolor*, announced by Heslop Harrison in the *Journal of Botany* of July 1941, conveniently provided strong confirming evidence for the periglacial theory. But its usefulness in supporting Heslop Harrison's pet theory might also have made it the straw (or sedge) that broke the camel's back for the British botanical community. With this 'discovery', Heslop Harrison had gone too far and had to be stopped. As it turned out, John Raven, son of the distinguished naturalist Charles Raven, was the man nominated to do it.

3

The Professor and the Don

In spite of a childhood spent reading Enid Blyton novels, I now realise that there is no universal law that says that likeable people are good and unlikeable people bad. Nor for that matter is there general agreement on what attracts us to some people and makes us shy away from others. At fifty years' remove, it is clearly dangerous to grub around in memoirs and obituaries, hoping to get near the nature and motives of the two men in this story. And it's doubly difficult with Heslop Harrison. Not only is there less written about him as a person than there is about Raven, he also leaves fewer surviving friends and relatives. And the closest relatives who were still alive at the time I began to explore the story resolutely refused to reply to my letters or to return phone calls.

It may be argued that, apart from the interest we all have in other people, there is no need to know about someone's personality, capacity for friendship, or emotional stability in order to decide whether or not he has committed some misdemeanour. Certainly, as I read Raven's account of his trip to Rum and of the powerful, but still circumstantial, case against Heslop Harrison, I was predisposed to believe it on the facts alone. But 'facts' may not turn out to be what they seem. Even the most persuasive case may fall apart when a few more facts are known. Our understanding can be conditioned by what we want to believe, and this could be as true of Raven as of anyone who reads what he wrote.

Sometimes it helps to know the nature of the messenger as well as that of the message. If what I could discover of Heslop Harrison's character was inconsistent with skulduggery, or if Raven's revealed a streak of malice, the story might not be quite what it seemed. And in either case, motive was an important factor. If there was a motive for Heslop Harrison to fake his results, the case against him might be strengthened, just as if there was a motive for Raven to besmirch Heslop Harrison, his case might be weakened. For that reason I began to look for clues to the characters of both men in what I read about them.

There were many contrasts between them in what I began to discover: 'His brusqueness and strong personality often brought him difficulties in human relationships'; 'His strength lay in a readiness to give himself, invoking a response in which we became vividly aware of ourselves, and in this self-giving he remained fully himself, so that our own projections and expectations could find no purchase on him.' The first was a colleague's description of Heslop Harrison; the second, a friend of John Raven's. And here are two more character vignettes: 'His intellectual gifts, his capacity for applying them to a wide variety of subjects, his expertise in his specialisms, together with his independence of outlook and his whole upbringing, all built up in him a superb self-confidence'; 'He was plagued by a sense of his own inadequacy . . . The symptoms of his self-doubt are everywhere discernible: in the acutely physical attacks of nerves before he lectured or after he had landed a salmon; in the overindulgence of alcohol which seemed at one time to be getting out of control; in the excessive smoking that persisted to the end of his life . . .' Again, first Heslop Harrison, then Raven.

Raven's redeeming feature for those who did not share his interest was a facility for conveying the passion, the fascination, and even the tedium of his quest in his descriptions of 'botanising', as it is called by its practitioners.

Raven did not write much about botany – at least not compared with Heslop Harrison – but what he wrote has an easy, invigorating style. He collaborated with Max Walters on a book published in 1956 called *Mountain Flowers*, one of a famous and much-collected series of natural history books called The New Naturalist. In a preface to a reprint in 1984, Walters drew attention to a characteristic of the book that strikes anyone who reads it: its two very different writing styles, which Walters ascribes to the professional-amateur difference between him and Raven, one adopting the scientific approach, the other the aesthetic. I'm not sure that that is the reason – the two men also had very different personalities – but the two styles combine well in alternating sections to convey both the scientific background of the topic and the human activity of botanising. Unlike Raven's professional writing on classical philosophy, there is a wayward and meandering feel to his chapters in *Mountain Flowers* that makes you feel as though you are listening to a man who has so many things to tell you that he doesn't know where to start.

Here, for example, is an extract from a section by Raven in a chapter called, soberly, 'The Cairngorms and Lochnagar'. He has just described a plant that is 'one of the few British mountain plants which demand both a certain agility and a fair head for heights'.

I cannot resist, in this connection, recounting the Strange Episode of the Major's Spectacles. One hot day in the middle of July I was descending alone one of the long scree gullies that drop into Caenlochan Glen from the west. When I was almost immediately below the ledge on which the sow-thistle grows, I noticed, lying on the scree, and somewhat mud-bespattered but otherwise intact, a pair of horn-rimmed glasses. Thinking abstractedly that their owner would never find them there, I picked them up, put them in my pocket and promptly forgot them. When eventually I emptied my pockets, I was at a loss. To take them to the Braemar police seemed too forlorn a hope, and no alternative occurred to me. In the end, cursing myself for not having left them where they lay, I threw them in disgust, along with the refuse from my vasculum,* into the waste-paperbasket in my hotel bedroom and left for Aviemore. Next day, accompanied by three fellow-botanists (only one of whom, incidentally, and that not the Major, was previously known to me) I paid my first visit to the hawkweeds of Glen Einich. The cliffs around the head of the loch are formidable and treacherous, and the hawkweeds have a bad habit of growing just out of reach. Suddenly, when we were each busy with our own ploys, a gentle voice reached me from along the cliff: 'Could somebody come and give me a hand?' Two of us arrived in time; the Major was still hanging on, and we grounded him safely. 'What a good thing I wasn't alone,' he said calmly. 'Last year I got stuck like that on a ledge in Caenlochan where the blue sow-thistle grows. I was very keen to photograph it and scrambled up to it without bothering to think how I should get down again. That time there was nobody to help, and in the end I pitched on to my head on the scree and was hard put to it to get back over Glas Maol to the Devil's Elbow.' 'You didn't by any chance lose a pair of horn-rimmed spectacles, did you?' I asked, and went on to tell him, somewhat disingenuously, that I had left them in the Invercauld Arms Hotel at Braemar. He wrote to the manager, of course, and was disgusted with the reply that nothing had been seen of his glasses. I never dared tell him the whole shameful truth.

Many of Raven's friends speak of his sense of fun and his ability to laugh at himself, and that passage – if you believe that the confessional quality was not contrived – is typical of Raven's approach to life.

---

* For more about vasculums (or vascula), see Chapter 4.

When I read similar accounts of botanising by Heslop Harrison, I was interested to see the difference between the writing styles of the two men. Where Raven was witty, Heslop Harrison was arch; where Raven was confident, Heslop Harrison was florid. Several of Heslop Harrison's pieces for the *Vasculum*, a local botanical journal, even when reporting serious botanical news, had the feel of a 'What I Did During the Hols' school essay. Here is an extract from one such piece:

> In the end, dragging ourselves from our labours, we returned to the lighthouse where, fresh from a really good wash, a welcome relief after the intense heat of the bare rock surfaces and the brilliant sun, we sat down to a 'scrumptious' tea and a chat with the keepers. Next, a microscope, presented to the lighthouse to afford recreation during the long dark nights, was produced and adjusted. However, a look at the clock gave us warning, and with many regrets we proceeded to the landing-stage. There, taking leave of our kindly hosts, we stepped aboard the boat.[10]

And two more quotes:

> In this pool were vast numbers of water beetles, chiefly *Agabus bipustulatus*, but their presence did not deter us from quaffing deeply of the water, or, in some cases, of preparing lemonade with it and the powder bought for that purpose from the 'very general' shop of Inverarish . . .[11]

> Our kindly hosts pressed on us fresh vegetables, berries of various kinds, new milk (a very welcome gift), jam and scones. Thus, when we left them, we had the knowledge that, if the inhabited portions of the island extended to only a very few acres, and the population consisted of only a dozen souls, all had

accepted us as part of themselves and showered on us kindnesses which grew in magnitude as our stay lengthened . . .

Admittedly, these are quotes from articles providing general descriptions of botanical trips, but even a paper with a more scientific purpose, '*Microcala filiformis* H. and L., a Flowering Plant New to Our Counties', tells us rather more than we really need to know:

> . . . on the afternoon of Monday, July 13th, it was necessary for me to take a school's examination at Blyth, and I had, of course, to prepare the necessary material. To make sure of the freshness of this, I arranged to collect it on the sand dunes between Seaton Sluice and Blyth, and, therefore, instead of taking a ticket to Blyth, I booked to Seaton Sluice and walked along the dunes . . .[12]

Without reading too much into the difference in writing styles, Heslop Harrison's writing – clunky and contrived when set against Raven's elegant and surefooted prose – feels like the writing of a man with a mission, keen to impress and to impose his personality on the reader. In exploring the motives and actions of these two men, their writings were the best, and sometimes the only, material I had to go on.

To start with Raven, his contributions to *Mountain Flowers* leave no doubt about his enthusiasm for botany and his detailed knowledge. Raven was particularly passionate about a group of plants called hawkweeds. 'There are,' he writes of one group of species, 'few British plants more strikingly handsome than these mountain hawkweeds, with their neat basal rosettes, their velvety-black involucres, often with a halo of long white-tipped silken hairs, and their broad golden flower-heads.'

He would go to great lengths to see unusual species of hawkweed that he had never seen before, conveying in his writing both the frustration of the quest and the tender delight when he succeeded. One particular hawkweed, which he had never seen, had been described by its discoverer, the Reverend H.E. Fox, as being 'on the back of Kirk Fell, Ennerdale.' But Raven wasn't sure where to look, since he didn't know which side of Kirk Fell it was that Fox saw as the back:

> I guessed wrong, and I spent two whole days (for which, however, I was rewarded by the discovery of the lovely *H. holosericeum* in a new locality) on a false trail. But the third day not only brought me, by elimination, to the right ghyll, it also revealed both the objects of my search; and a very thorough exploration of the whole ghyll showed that, while there are now only three plants of *H. praetenerum*, of *H. sparsifolium* there are still only two. In all my days on the hills I do not think that anything has struck me as more remarkable than that these two very rare plants, in the passage of sixty-five years, should neither have disappeared nor appreciably increased, but should rather have remained almost exactly as they were. When, as is probably the case here, a plant is a diminishing species, the actual process of diminution is evidently sometimes very protracted.

So far, in fact, it all sounds like little more than going for long, bracing walks to look at the scenery and 'tick off' a few plants on a list. But the science of botany, particularly the part of it that deals with the distribution of species, relies on meticulous and informed observation of plants from just such people as Raven, who are prepared to tramp the countryside and bring back data in a very systematic way. Everything Raven observed he noted on small cards, often sealed and kept away from prying eyes if they contained specific information about the location of rare plants. Over the years, pursuing the hobby of natural history turned many amateurs into essential providers of information to the professionals. Finding a plant and bringing it back was one thing, but information about where it lived – the soil, the climate, the other plants around it – was as important as the plant itself in helping to place it in the big picture.

Mountain Flowers was published some years after Raven wrote his report on Heslop Harrison and dealt directly with many of the issues that underlay Heslop Harrison's work and his area of research. But no mention was made of Raven's findings, although there is one rather prophetic reference to an earlier botanist, as distinguished as Heslop Harrison, who wrote the first book published in Great Britain that was dedicated to one genus only. And that genus was Raven's passion: the hawkweed. The monograph's author was James Backhouse, and Raven was scathing about the quality of his botanical statements:

> Unfortunately (or, as some may think, wisely), as soon as the Monograph was written, Backhouse himself seems to have lost interest in the genus. Although he continued for over thirty years to make authoritative pronouncements on the specimens referred to him, his utterances bore a steadily diminishing relation to reality. But until his death in 1890 there seems to have been nobody who was prepared, on this subject at least, to gainsay him.

This was, of course, written while Heslop Harrison was still alive, aged seventy-five. And nobody, even Raven, had been prepared to gainsay *him* in public.

Although both men would consider themselves serious botanists, Raven knew that he was not a scientist. He had no scientific training or degrees – not that he saw that as any sort of lack – even though with his father, Canon Charles Raven, his childhood had been so dominated by natural history that observing and classifying plants and insects had become a way of life. Between them father and son set out to paint exact and faithful pictures of every single

British wildflower. It was an extraordinarily ambitious task and its fruits are visible today in a long row of ring binders kept in cupboards in Faith Raven's house near Cambridge, and in a book published in 2012 of beautiful flower paintings by John and his father. A correspondence between John and Charles Raven in the 1940s shows how passionately and single-mindedly John went about the task, and how fulfilling it was to him, as he scoured the country in all weathers, ticking off a master list and revelling in the growing number of plants gathered, which he would send to his father to be painted: 'The snow having gone, my longing for the beginning of the season is growing. And my hopes for a good bag are very high' (February 5, 1945).

Raven's longing led him to think about how to satisfy it, and later that year, two weeks after the end of the war with Germany and on top of a London bus, he made a decision:

My dear Pa,

I've been thinking deeply, with the following surprising results: –

My departure from Oxford House* at the end of next month marks the end of an era and the beginning of quite a different one. I shall leave here – as nearly everybody in the world is at present – somewhat jaded, and – unlike most other people – temporarily devoid of conscience. Before starting on the next job, whatever that proves to be, I would like to take a really good holiday and incidentally make myself really fit. And so I spent half an hour on top of a bus this morning concocting the following scheme.

When I leave you on Iona – say on July 14th – I propose to remove your bicycle and, using it and such trains as are available,

---

* The East End charity based in Bethnal Green with which John Raven worked after leaving university. See page 37.

move slowly across central Scotland to the East Coast. My proposed itinerary and various objectives are as follows:

14th–19th at Dalmally, whence I shall explore Ben Laoigh for (in particular) *Cystopteris montana*.

19th–21st, to the Lawers Hotel, whence I shall go for *Carex microglochin* . . .

21st–24th, to Pitlochry, whence I shall go for *Menziesia, Carex rupestris*, and *Schoenus ferrugineus*.

24th, meet Peter* and proceed with him to Braemar . . .

Assuming that I can find a room somewhere in or near the places named, and that it occasionally stops raining, I cannot really see that I should have to be excessively energetic to get almost all these treasures. I should, of course, make careful inquiries in advance. Peter is keen on his share in the exploit, and for my part, though I feel it must be immoral to contemplate anything so exciting, I am not at all sure really that it isn't a pretty brilliant scheme!

But of course there is one big snag, namely funds. Even though Peter will be paying for himself, my own expenditure, by the time I'm back in the South, can hardly be much under £20; and as you know, all I possess at present is a few bad debts. I assume that one day – though it may not be for about 2 years – I shall get a job where I shall have a penny or two to spare. The question is whether, until that indefinite date, the family coffers can rise to lending me so substantial a sum. Would you please give me an honest and considered answer to this question? And if by any chance it is yes, then I will get down at once to the problem of booking rooms . . .

If you think this whole project too absurd, then say so. It's a superb day and the prison house of B.G. (Bethnal Green) sends

---

* Peter Kuenstler, a friend from Oxford House, who describes himself as 'playing a kind of bumbling Dr Watson's role to John's botanical sleuthing.'

my fancy roaming on fantastic expeditions. But all the same I'd like to do it.

　　With much love,

　　B.*

The munificent sum of twenty pounds was produced from 'the family coffers' and Raven set off, sending back to his father regular reports and parcels of plants:

> Herewith some beautiful specimens of *T. stellatum* found at 11.15 this morning at Shoreham. If you have the energy, paint the 3-branched plant. Also *Vicia lutea* found at the same time and place . . .
>
> 　　Also some good material *Lathyrus hirsutus*. There was a lot of it, but none in flower. I suggest that you paint the whole of the relatively short piece that is nearest to flowering except the actual flower, then try and force it out. If you fail I will try and get down there again in a fortnight's time to paint the flower – and nothing else – on the spot . . . (29 May 1945)
>
> P.S. What do you make the season's total now? (16 June 1945)
> The holiday's total stands at 12 and the year's at 29. We'll reach 50 yet. (15 July 1945)

In spite of being 'temporarily devoid of conscience', John Raven occasionally recognised ethical dilemmas even in the field: '. . . when I had concluded that it was moral to dig a bit up . . .' (July 26, 1945).

Max Walters was impressed by Raven's knowledge of the exact location of the rarest British plants: 'He had an instinct for a "good" site. He used, almost subconsciously, signs of "indicator species" – commoner mountain plants which were sufficiently choosy about

---

* Raven was known as Bunny among his family.

where they grew – to mark off the site as worth searching for the real rarities, but also, one had to admit, he seemed to have an inexplicable skill in prediction. When this was combined, as it so often was, with his capacity for very painstaking detective work, the combination was very powerful indeed.'

However you classify Raven's botanising, the rigour he applied to noting, observing, and classifying plants had all the elements of scientific observation, and even the passion he displayed in his pursuit of Heslop Harrison for what he saw as the betrayal of science was worthy of any committed professional scientist.

But was there more to it than that? Did he embark on his 'painstaking detective work' to investigate Heslop Harrison for some other reason? Did he bear a grudge against Heslop Harrison or not take him seriously as a botanist?

Faith Raven said: 'I don't think he would have felt vindictive [toward Heslop Harrison]. I don't think he would have pursued him for the sake of his own vanity. Although he was a very keen field botanist, it was not his career, and if John lacked anything, it was ambition. I think he more loved the chase and could be aggressive in an inverted way. I don't think he would have been vindictive towards a total stranger.'

But he could be unkind to people he didn't like: 'What people said was that if you in some way annoyed him, you could be dropped. There was a man who was a very close friend of his called David. When David married a woman who was madly interested in horses I think John really never spoke to him again. It sounded as if you *could* put your foot in it and then you were dropped, but you certainly wouldn't be pursued.'

He could also be aggressive. 'There was a day when I was expecting the twins,' Faith Raven said, 'and we'd got our local doctor and his wife for dinner and the headmaster of the King's College Choir School. The College Council had just that day refused to make a contribution to a new assembly hall for the school and John said the

College were absolutely right. Although our boys were already at the school, he thought that this was special education and very favoured. I did say afterwards, "Well, if you are going to be rude to the guests – which you very seldom are – do tell me beforehand and then we'll delay the dinner party, particularly when I'm pregnant." '

It could be that Raven, as a well-bred Cambridge man, felt a natural antipathy to someone of Heslop Harrison's social background. But there is plenty of evidence that Raven was not a man to whom social class was a preoccupation. As a conscientious objector, he had spent some time during the Second World War doing social work in Oxford House, a settlement in the East End of London. Like Cambridge House and Toynbee Hall, this was an institution founded in the nineteenth century in the deprived areas of London to enable graduates to carry out social work among the disadvantaged. Here, Raven organised places in schools outside London for children who were evacuated because of the German bombing.

When he left Oxford House after the war, he wasn't sure what he was going to do. Then he was offered what he described in a letter to his father as 'the most tempting job in the whole world', a research fellowship at Trinity. In this case the subject was classics, but Raven's excitement at the opportunity reflects what was probably his true motivation in pursuing Heslop Harrison: a passion for knowledge and a fierce belief in the purity of the intellectual pursuit, whether applied to Plato or plants.

Raven lived up to the public image of an Oxbridge don, with his intellectual passions, his donnish wit, and his playful eccentricities, almost as if he was afraid to be seen to be too ordinary. But he could also arouse the suspicions of the more conventional dons. In 1948, he was offered an even more tempting job, to teach classics at King's.

'When John arrived at King's,' Faith Raven said, 'they thought, "This man will be rushing round in shorts. He's been doing social work in the East End, and he's interested in the Scottish mountains and botany – he'll be an absolute nightmare." But they found he was able to drink with the next man and had a sense of humour and played bridge, and so they became very fond of him.'

If Raven was 'plagued by a profound sense of his own inadequacy', he could disguise his defects, real or imagined, by behaviour that made an immediate impact on people and led them to appreciate his surface qualities without probing too far into his depths. Stories of travelling in the luggage rack of trains, applauding a piano performance so enthusiastically with a glass of port in his hand that he fell into a coal bucket without spilling a drop, and even the cross-legged party trick I observed as an undergraduate, all show a very human tendency to put on an act as a means of hiding what he saw as his real self.

So perhaps he put his real self into his botany. Faith Raven wrote, 'His most intense experiences came during interaction with plants.' A friend, Tom Creighton, who was to share some of the Rum experiences, said, 'He admitted to me that he had a mystical feeling for vegetation, and I think this grew stronger all through his life.' And along with this mystical feeling went a strong sense of ethics.

In his statement before the tribunal he faced as a conscientious objector, Raven said: 'One of my strongest moral beliefs has, for some years past, been that no good can be achieved by evil means.' Even if Heslop Harrison thought that he was doing good by bolstering a theory he believed to be true, for Raven this would be a heinous crime. Tim Clutton-Brock told me that Raven was 'tremendously interested in people, terribly sensitive, hypersensitive of people's feelings.' But when I pointed out that his hypersensitivity to people's feelings didn't seem to have extended to Heslop Harrison, Clutton-Brock said, 'I guess Heslop Harrison, from John's point of view, had stepped outside the pale.' And the pale in this case was the boundary that separated truth from falsehood, good from evil.

In embarking on his quest, there might have been one factor that made it easier for him to do so unnoticed. As Clutton-Brock put it, 'Heslop Harrison was an established scientist, and John was an amateur flower student, so Heslop Harrison hardly reckoned that John was worth worrying about.'

This is not really surprising. After all, at the time of the Rum affair, Heslop Harrison was a professor and Raven not much more than a boy (as a sixty-six-year-old might have said of a thirty-three-year-old). It was a severe misjudgement but not a surprising one. However many experiments he carried out and observations he made, I get the impression that, in the end, Heslop Harrison believed that the most important reason for people to believe what he said was because he had said it. By the late 1940s his career was established, his public reputation was firm, and he had daily reminders of his own achievements in a long string of degrees and publications. Unlike Raven, he saw himself as a scientist, first and foremost: distinction in his chemistry finals; scientific papers on genetics and entomology; government research award to the University of Durham; lectureship, readership, professorship, and fellowship of the Royal Society at the age of forty-seven. His knowledge of plants and insects was all-encompassing.

From the library of the Royal Society, Britain's prestigious club for distinguished scientists, you can buy a slim grey pamphlet containing an obituary of Heslop Harrison. No dead fellow of the Royal Society goes without an obituary. Its absence would draw attention to activities suggesting a severe misjudgement on the part of all the fellows in electing this man or woman in the first place.

So I walked into the marble halls of the society, just off Carlton House Terrace in London, and asked my way to the library, where an archivist handed me a copy of the professor's obituary in exchange for £3.50. Not a bestseller, I thought, as I flicked through its twenty-six pages of elegantly printed but seriously scientific stuff. When I sat down to read it properly I discovered that it contained a surprising amount of useful corroboration of the picture I was forming of the professor, conveyed not least by its air – it seemed to me – of damning with faint praise. In fact, quite often it damned with faint damns.

Heslop Harrison's obituary was written by Professor A.D. Peacock (another society fellow, by custom), a friend of Heslop Harrison's from University College, Dundee. Much of the booklet was fairly straightforward stuff – early career, degrees and awards, key scientific papers, lectureships and professorships. But Peacock made a fair stab at summarising the personality of his friend:

> Harrison was a striking figure; tall, slightly stooping (the 'search-ing posture of the field naturalist'?), black suit and tie; black hair and neat black moustache; above all, vivid grey eyes in a pale face. Learning came easily to him. He was quick of apprehension, had a prodigious memory, and an encyclopaedic knowledge of his specialisms. He could surprise one by his wide general knowl-edge, being particularly well versed in English literature from Chaucer to Dickens, and very knowledgeable in world history and politics. His grasp of geology and geography is obvious from his biogeographical work. Equally he could surprise one by his near-philistinism regarding modern aspects of literature, poetry, art, and music: in fact he enjoyed stuff like 'Sexton Blake'.*

But all the way through the obituary, the bland bonhomie of a fellow seeker after truth – *nil nisi* and all that – seemed to be punc-tuated by sudden shafts of doubt, ambiguity, and occasional open criticisms.

> Harrison's experiences in his boyhood schools and his workaday overgrown village, rife with parochialisms, were rough and

---

* As the grandson of a writer of Sexton Blake stories I bridle a little at this.

tough, but they developed his capacities for holding his own and for assuming leadership: they left their stamp upon him for life ... His brusqueness and strong personality often brought him difficulties in human relationships ... Rooted in his native Birtley, he laughed at pretension, patronage, and privilege; excellent company though he could be amid friends of diverse interests and social classes, and among fellow-biologists, he refrained from making certain professional contacts at home and abroad and so ran the risk of inviting misunderstanding and criticism.

Clearly, as I was beginning to find, the invitation was sometimes accepted.

Peacock's friendship didn't stop him remarking on those aspects of Heslop Harrison's trenchant writing style that might have led to this 'misunderstanding and criticism':

... at times, his strength of feeling brought a style over-emphatic and insufficiently impersonal. Of the latter here are some random examples. With Harrison a case is of 'paramount importance', a result is 'absolutely decisive' or 'without a loophole for error', there can be 'no escape from the conclusion', and argument 'confutes this entirely' or is advanced 'to clinch the matter', one is 'compelled to admit' and must note that 'the outcome, to say the least, was astounding'.

But Peacock also quoted warm and obviously genuine tributes to Heslop Harrison's prowess in the field, including the following from one former student:

I very willingly relate some of my experiences with Harrison at home and abroad. In his native north country he had certain haunts, especially if they were places in the midst of industrial surroundings. He did a number of things at the one time. Not content with finding and naming specimens he must needs see if they were freaks or hybrids. On one famous excursion on the fells the party marvelled at the way moths came steadily to him as if he were the Pied Piper. He had a female Eggar in his pocket and its scent was causing the assembly of males. He was never pedantic. I know how he held a group of children fascinated by his talking about willows. He was especially fond of Waldridge Fell, a little hidden place primaeval in some ways. I once went with him and an amateur coleopterist to the south of France and on the dunes near Cette [probably Sète] the coleopterist lamented the absence of beetles. Harrison pulled up the grasses and succulent plants and there were the beetles lurking underground. On another occasion I accompanied him on a trip to the Pyrenees and I have an abiding memory of a place he showed me where he had recorded the number of species found: I call it the Little Valley of Sixty Kinds of Butterflies. I forget fifty-nine of these but will never forget the glorious Apollos flighting downhill into the hot sunshine.

There was clearly a quality of showmanship about Heslop Harrison. I get the sense from such descriptions that he was no solitary stalker of the hills, communing with nature for the sheer joy of it; he required – or at least enjoyed – an admiring audience. Max Walters told me of a field trip he went on with Heslop Harrison and a group of botanists some years after Raven's investigations. Even though what he said was tinged with hindsight, he painted a picture that was consistent with Peacock's obituary account but also suggested a major egotistical streak in his botanising: 'There were twenty people who were obviously rather under his spell. He was entirely in charge of the thing. He was a good botanist and seemed to know his plants well. But if you contradicted him it was dangerous, and therefore I didn't contradict him.

He was pontificating about roses, standing by the bush and saying, "This is so and so." It was fairly obvious it wasn't the sort of situation where you could even throw a minimum doubt on whatever was going on. Certainly I came away with the idea that he was a guru for a small group of people, and he behaved like that. It's not unique – I could even do it myself if I wasn't careful, you know, pontificate about things and say, "This is so and so," and be rather thrown if some bright young student said, "Are you sure?" '

In the Royal Society obituary, one of Heslop Harrison's colleagues praises his expertise:

> If slight misquotation be allowed, it could be said, 'The range of the mountains and islands were his pastures and he searched after every green and winged thing.' To novices, his powers of observation and knowledge of plant and insect life savoured of magic; professionals, too, could not forbear to cheer. Yet the explanation of his expertise is simple enough: he had a flair, together with long experience and the habit of 'doing his homework' if he planned to explore some distant region at home or abroad. In short, he knew the questions to ask.

One person who fell under Heslop Harrison's spell was John Morton, who, as a schoolboy in Sunderland during the Second World War, was a keen naturalist with an interest in flowers, birds, butterflies, moths, beetles, and shells. I felt from the beginning of my researches that I needed to talk to someone who had known Heslop Harrison personally, and Morton had had a long association with him and helped me understand something of the professor's complex personality. 'I came to know Prof. Harrison,' Morton said, 'when I attended meetings of the Northern Naturalist Union where he was a leading light. He impressed me greatly. He had an imposing presence – a very tall man, with a powerful voice and a commanding personality. Also, he was a man of remarkable

...wledge and ability. He went out of his way to help young naturalists like myself and spent much of his time with them in field
trips. On the other hand, he could be devastating in his response to
people who questioned his statements or disagreed with him.'[13]

At the end of the war, Morton applied to study at King's College,
Newcastle upon Tyne. Heslop Harrison was on the board that
interviewed him, and he was accepted for a degree in botany.
Morton eventually lived across the road from Heslop Harrison in
Birtley and was to see a lot of him in the years after he retired.
Although he accepted a position at an African university after
completing his Ph.D., Morton returned each year to England for
three months and stayed in Birtley, where he kept up his friendship
with the professor.

Clearly, Heslop Harrison had a gift for inspiring the young, and
Morton displayed a lifelong gratitude for the skills he had been
taught and the science he had learned. Unlike Raven, Heslop
Harrison was trained as a scientist and could therefore use his
observations to build theories that would lead in turn to more
observations. Observing is perhaps the first stage in the scientific
enterprise, and asking the right questions is the second, but perhaps
the most important stage is the process that provides the answers:
the scientific method. Rigour and skill in using the scientific
method will yield reliable inferences from observations. But some
of Heslop Harrison's critics saw his research style as slipshod. Even
Morton accepts Heslop Harrison's defects in this area: 'A major
shortcoming as a scientist, in my view, was his failure to keep good
and properly documented specimens of the plants and insects that
he found and the ones that he worked on. He just did not have the
patience to spend the necessary time to do this essential part of
research. As a result, he left himself open to the criticisms of those
who questioned his discoveries. The adequate documentation of
those discoveries in many cases just isn't there, and so the criticisms
easily stick and will probably go on forever.'

Throughout his professional life, Heslop Harrison carried out experiments on plants and insects, using individual specimens, dead and alive. As the practice of science demands, he should have published the results of those experiments in such a way that his colleagues could check his results. But, unusually, he would often carry out his research at home, an environment that, from one description, sounds less than ideal:

> . . . in the study, bathroom, kitchen, tool-shed and other situations where living conditions were appropriate, he distributed the stages of his insects in glass-lidded tins or in simple breeding cages of his own design and make. Only thus could he give the insects unremitting care and be on the spot at critical times during life-histories and when pairings had to be effected. (The forbearance of his household may be imagined.)[14]

There's something odd about this image, at least as the basis for serious scientific research. First, bathroom, kitchen, and tool shed are surely not environments as good for carrying out long-term insect-breeding research as a purpose-built biology laboratory. Presumably from time to time people – members of the family – must have bathed, cooked, and hammered in these various environments, unless the forbearance to which the obituarist Peacock refers implies the exclusion of the family from these activities when *Tephrosia bistortata* was at a sensitive time in its reproductive cycle. Second, unlike in a lab, there would have been no lab assistants, technicians, or academic colleagues to look over his shoulder, discuss whether the moths were really doing what they were meant to do, or offer counter-theories to some of the more unlikely hypotheses Heslop Harrison put forward. And then there was the garden:

> His garden was used only for scientific work. It was on a slope facing south but was no showplace, for all its plant life served

some useful purpose – as a place of rose and willow hybrids, as foods for insect larvae, as shelter for pots of pupae, and as hosts for insect galls. One entomological friend remarked that its spirit reminded him of L'Harmas, Fabre's[15] wilderness in his small estate in Serignan where nothing was disturbed unnecessarily. The greenhouse was used for crossing experiments with primroses, for rearing *Melandrium* for the study of sex chromosomes, and for plant breeding.

With a distinguished professor, hugely confident about his own way of doing things, and a young academic possessed of strong moral sense spiced with an appreciation of fun, the stage was set. And there was one other factor that was to make the confrontation explosive: Heslop Harrison's capacity for seeing enemies around him and sometimes for creating them. In Morton's words, 'The Prof had many enemies because he was so dogmatic and outspoken, and devastating in his criticism of other people.' Morton – Heslop Harrison's friend, remember – went on to reveal that Heslop Harrison was not, as he was to assert often, surrounded by a horde of loving and respectful colleagues: 'In particular, he had earned the enmity of some people in his own department at Newcastle. He ruled this department as a despot.'

Heslop Harrison's dogmatism was a reminder of the attitudes of an earlier era, described by Bertrand Russell in 1931: 'When it was first proposed to establish laboratories at Cambridge, Todhunter, the mathematician, objected that it was unnecessary for students to see experiments performed, since the results could be vouched for by their teachers, all of them of the highest character, and many of them clergymen of the Church of England.'[16]

When dogmatism and despotism are wielded as weapons, the injured parties cannot be blamed for using any opportunity to get back at their attackers. When I asked John Morton why so many botanists and other natural historians came to believe the rumours that Heslop Harrison had faked his results, he said, 'His enemies must have jumped at the opportunity to get even with the man who had bested them in so many bitter arguments and all too often humiliated them with his scathing comments and criticisms. The Prof was admired by many but loved by few.'

As John Raven embarked on his task, he had three goals: to find evidence of Heslop Harrison's misdeeds; to persuade Heslop Harrison that he'd been found out; and to publicise proof of the dubious nature of some of Heslop Harrison's research. It was to turn out, unexpectedly, that the first was easier to achieve than the second or third.

4

Rumours

I wanted to learn more about the world of British botany in the 1930s and 1940s, and my opportunity to do that came unexpectedly one day in the London Library, a treasure trove of out-of-the-way or hard-to-find books, most of which can be borrowed. All the shelves are open and so instead of looking up a book in a catalogue and asking someone to get it for you, you can roam the shelves to your heart's content. In fact, you often find yourself roaming the shelves to your heart's *dis*content because the nineteenth-century system of classifying books is often impenetrable to twenty-first-century logic, and the system of locating the books once classified requires the geographical skills of a Burton or a Speke. But the illogical and labyrinthine nature of the place is such that, if you don't find what you are looking for, you almost always find something better.

One day, I attempted to find the periodicals section of the library where past issues were kept, to look through some back numbers of the *Journal of Contemporary History*. I had tried to do this once before, but I gave up after being confused by a notice that appeared to direct me through a solid wall (or it might have been upward through the ceiling). This time I persevered and was successful. I found the *Journal of Contemporary History* and, not unexpectedly, it didn't have what I was looking for, but two shelves below were bound volumes of the *Journal of Botany*.

Now, I would normally never have thought of looking in the London Library for specialist botanical journals: it is known much more for the quality of its humanities collection – literature, history, art, biography, topography, and travel. Its science section is a ragbag. This is made clear by the fact that one of the largest sections of the library is labelled 'SCIENCE' and is actually the place they put all the books they can't classify under the other broad headings. The list of categories is arranged alphabetically, so that you find evocative sequences such as 'Fashion, Fear, Flagellation, Flax' and 'Electroplating, Elephants'. 'Inns of Court' is juxtaposed with 'Insanity', and 'Sex' is between 'Sewage Disposal' and 'Sheep'.

So far in my researches the only access I had to Heslop Harrison's original reports from Rum was in reprints that were included in the Raven file. Some of these reports were first published in the *Journal of Botany*, although I didn't have a note of the references with me. The bottom shelf of volumes stopped at Volume LXXIX, the volume for 1941, and when I looked at the top of the next shelf, the section moved on to another periodical. So I pulled out Volume LXXIX of the *Journal of Botany – British and Foreign, edited by J. Ramsbottom, O.B.E., DR. SC., M.A., V.– P.L.S., Keeper, Department of Botany, British Museum (Nat. Hist.)* and, leaning against the shelves, began to read.

I was expecting a series of abstruse and incomprehensible (to me) papers on botanical topics, and I found them. But I also got a more rounded picture of the preoccupations of botanists set in the context of the beginning of the Second World War.

For example, in July, the editor of the *Journal* thought it worth drawing the attention of British readers to concerns expressed in America by the president of the Rockefeller Foundation about the Nazi threat – to botanists:

In the shadows that are deepening over Europe the lights of learning are fading one by one . . . As German forces have moved into

country after another a definite pattern has been followed in
relation to the universities and other schools . . . the measures of
repression adopted by the occupying authorities included the
closing of the institutions, sending faculties to concentration
camps and even breaking up student demonstrations with machine
guns and tanks . . . The condition of university life and standards
on the Continent is now little short of appalling, but we have still
to learn of the sufferings of other botanical workers . . .

Among the occasional news of the deaths of botanists – 'We
deeply regret to record the death of Mr Francis Druce through
enemy action' – was equally distressing news of the destruction of
plant life by German bombs:

During a recent air attack the whole of the glass in the rooms of
the Linnaean Society, Burlington House, was smashed, but only
trivial damage was otherwise sustained . . . The South London
Botanical Institute recently received the contents of a Molotoff
bread-basket and the resulting fire caused considerable damage
to the building. The herbarium suffered though not greatly . . .
The Chelsea Physic Gardens . . . has suffered considerable
damage, chiefly to the plant houses . . .

And, after a decent interval to foil Nazi subscribers to the *Journal*:
'It is now permissible to reveal that the British Museum (Natural
History) has suffered damage from air attacks. The Department of
Botany has been most seriously affected, the general Herbarium
having been set on fire.'

Another sign of the effects of the war on the practice of botany
came in a note about efforts to replace imports of medicinal plants,
principally foxgloves, with the native version by mobilising the Boy
Scouts and Girl Guides Associations and the National Federation of
Women's Institutes to scour the countryside.

Alongside these hints of events in the wider world, the *Journal* did not neglect the finer points of botanical exegesis. There seemed an unusual preoccupation with the correct dates of certain minor botanical texts:

> The title-page of the first volume of Augustin Pyramus De Candolle's 'Regni vegetabilis Systema naturale' (octavo: Paris) is dated 'M.DCCC.XVIII' . . . There is, however, clear evidence that this volume appeared late in 1817 . . .

> Some authors have taken 1832, others 1831, 1833, 1834 or 1835 as the year of publication of the important paper by Alexander von Bunge (1803–1890) on the flora of northern China entitled 'Enumeratio Plantarum quas in China boreali collegit Dr. Al. Bunge, Anno 1831.'

And the most obscure:

> Demetri Brandza's 'Prodromul Florei Romane sau Enumeratiunia Plantelor pana asta-di cunoscute in Moldova si valachia'* is dated '1879–1883' on its title page. Contemporary references reveal that it was published in two parts, the first being dated '1879' but probably published in 1880, and the second published in 1883, as follows: – . . .

So obscure are these various paragraphs that the idea occurred to me for a moment that they might have been planted by MI6 to convey coded messages to our chaps in hiding in war-torn Europe.

As a reminder that even the humblest mould was at that time seen as a subject of botanical interest,† the following perceptive

---

* I have omitted this reference's scattered Romanian diacritical marks.

† In recent years, moulds and fungi have been reclassified to remove them from the plant kingdom and place them somewhat nearer to animals.

prediction was made, after an account of recent work on the mould *Penicillium notatum*:

> At present the supply of penicillin is small, so much so that the large amount consumed by intravenous administration was partly retrieved from the urine of the patients to be used again – a strange reminder of the Siberian method of providing continued religious exhilaration from *Amanita muscaria*. It would be remarkable if the one species of *Penicillium* proved to be the only mould capable of providing such an antiseptic, and we are doubtless at the threshold of a new era in antisepsis.

All of this browsing, which continued in the comfort of my study after I had established that Volume LXXIX was full of promise, gave a touching insight into how, even in wartime, the science and practice of botany must not be neglected. But it did more than that. It also gave me a glimpse of a rather stormier world than I had imagined and, in particular, of the degree of antagonism between botanists that occasionally came to the surface. A matter of naturalists red in tooth and claw.

Two prominent figures in British botanical circles from the 1920s to the 1950s were A.J. Wilmott and Maybud Campbell, who turned out to play an important role in the Heslop Harrison story. Wilmott worked at the British Museum's Department of Botany and rose to be head of the British Herbarium. He was very interested in the Hebridean flora and decidedly miffed that he had to share the territory with Heslop Harrison. The botanically named Miss Maybud Campbell was secretary of the Botanical Society of the British Isles (BSBI). Of all the people who are no longer alive to tell their story, Maybud, as she was universally known, is the greatest loss. She would have spilled the beans, I'm sure. According to Max Walters: 'The famous Maybud was a rich and in some ways spoiled lady. She was very intelligent, but she could be very naughty, and so she was

quite a problem in the BSBI. When she was nice she was ve
nice, but when she was bad she was awful.' When Walters s
I made a note to explore further the naughtiness of Maybu , ...u
my opportunity came a couple of months later when I met David
Allen, a man with an astonishingly high boredom threshold that has
allowed him over the years to sift through tedious botanical tomes,
tracts, journals, and the minutes of various natural history organisa-
tions and write rather racy and amusing books with titles like *The
Great British Fern Craze*.

Allen was a military-looking man in his early sixties with bris-
tling pepper-and-salt eyebrows and a white moustache. He was the
doyen of British botanical historians (in fact he may have been the
only British botanical historian), and so devoted was he to the task
that he even wrote a history of the Botanical Society of the British
Isles, a labour of love if ever there was one. I was eager to meet him
because he would clearly be able to evoke a period that seems to
have departed with a whole generation of botanists – during which
time the Rum affair took place.

I met Allen at a stage in my researches when, in occasional
moments, I wondered whether what Heslop Harrison did on the
Isle of Rum and what John Raven *thought* he did were of any
significance at all. Certainly many younger botanists had never
heard of the events – and 'younger' means under fifty – and there
was so little that I could find in print about them that it seemed as
if no one really cared whether or not the odd sedge or rush had
been misidentified or planted, or the result of a practical joke.

Any doubts disappeared when I met David Allen. As we sat
down to lunch in the cafeteria of the Wellcome Institute for the
History of Medicine, almost his first words were, 'Of course, this
was the greatest scandal of twentieth-century botany.' Because of his
knowledge of the BSBI and, indeed, his personal acquaintance
with many of the botanical experts of the time, he left me in no
doubt that during the late 1940s and the 1950s the profession had

been swept by rumours and anecdotes about Heslop Harrison's deeds, not just on Rum but throughout the Hebrides and even nearer home, on Teesside. Allen called Wilmott and Maybud 'the Ferdinand and Isabella of British botany', and it was from Allen's book and conversations that I learned of the rivalry that set Wilmott and Maybud against Heslop Harrison, as they each vied to establish a reputation as the definers of Hebridean flora.

According to Maybud's obituary in the botanical journal *Watsonia*: 'It was in 1933, through Gertrude Foggitt, that [to use her own words] "I penetrated behind the locked doors" at the B.M. [N.H.] and received the Keeper's permission to start work on *Salicornia* under A.J. Wilmott in the British Herbarium.'[17]

Colourful characterisations flew from Allen's mouth as he described Wilmott as 'touchy and unbalanced' and Maybud as 'intensely jealous of other women'. That she might also have blown hot and cold in her relationship with Wilmott was illustrated by Allen's story, told to him by a mild-mannered assistant curator – let's call him Barton – who described Maybud rushing into his greenhouse shouting, 'Mr Barton, protect me from Mr Wilmott!'

Allen's book on the BSBI, *The Botanists*, includes photos of Maybud and Wilmott, chastely separated at the top and bottom of the same page. Maybud stands in sensible shoes and calf length kilt on some rock-strewn Scottish hillside, looking coyly into the distance as if unaware of the camera. Beneath her is a picture of Wilmott, with little evidence of his touchy lack of balance, standing next to a tripod with a small, rather unimpressive camera on it. He is wearing plus fours and thick socks, and his camera is pointed at the flat cap of another botanist, Francis Druce – he who was later killed by enemy action – who seems in the photo about to fall into a stream.

In the archives of the Natural History Museum, Wilmott's correspondence is preserved in smart green files with his name embossed in gold on the spine.[18] I spent a morning trawling through it, looking for any letters that might throw light on him or some of the other characters in the Heslop Harrison affair.

There was an interesting collection of letters between Wilmott and Maybud, starting in a business-like and botanical manner and warming up as their acquaintance ripened. The correspondence starts in 1933 with Maybud addressing her colleague as 'Dear Mr Wilmott'. Then, in the mid-1930s, he becomes, rather provocatively, 'Dear Wilm'. At the same time, in his letters, she moves from being 'Miss Campbell' to 'Maybud'. After a brief cooling in the epistolary relationship in 1939 – when Wilm reverts to Mr Wilmott – the correspondence settles into 'Dear W' and 'Dear MSC'.

Much of the correspondence is purely about botanical matters, but occasionally a more informal note is struck by Maybud: on October 20, 1939, 'There's a most diverting squirrel just outside my window – hope it doesn't go to sleep too soon, I shall miss it! – a sweet grey fluffy cat has just sprung in and out of view . . . Yours industriously . . .' Wilmott's reply, written by return of post, makes only a tiny concession to Maybud's coyness: 'No squirrels here!'

A later letter to Wilmott reveals another side to Maybud: 'No "news" from BBC yet,' she wrote, 'beyond a note to say my application is receiving consideration. I got through my ENSA [a British forces entertainment organisation] audition (on Drury Lane stage) very well and am to start doing concert work for them in November for allied welfare.' Images of Maybud Campbell as ventriloquist, conjuror, or impressionist came to mind, but the truth was more upmarket. According to an obituary, 'One of Maybud's many attributes was a fine mezzo-soprano voice. In the 1930s she sang at occasional concerts in the Wigmore Hall. She specialised in Czech lieder, and decided to make use of this talent after the outbreak of war in 1939 by joining ENSA, where she was much in demand in places where the Free Czech Forces were stationed.'

David Allen's account of Maybud's time as general secretary of the BSBI is full of barbed references to her extravagance, tantrums, and schemes. She had a 'fondness for grand gestures characteristic of the concert singer that she had been till lately,' as well as a 'background of comfortable affluence (which led her ... to charge up taxis unthinkingly).' On the other hand, said Allen, 'she was by no means alone in those prewar days in displaying difficult behaviour: a generation still accustomed to flounce out of grocers' shops if not served promptly had far less patience with the frustrations of democratic procedures than those who have followed since.' Allen described a committee that had been set up to monitor threats to rare wildflowers by the encroachment of post-war development, and in an aside he said, 'Known as the Threats Committee, its meetings largely consisted of threats by the general secretary to resign.'

Judging by some of the things that went on in botanical circles, Maybud's grand gestures were occasionally entirely appropriate to the situation. Allen wrote of a notorious event at the annual meeting of the BSBI held in 1949 in Taunton, when Maybud insisted on a letter of reproof being written by the president of the BSBI to two rebellious members. Allen told me that the field meetings were often invaded by members of the Wild Flower Society, rather like bird-watchers in ornithology, whose only interest was in seeing and ticking off as many flowers as they could. On this occasion, two women who were determined to see some rarity had hijacked a bus to take them there, in the face of protests by other passengers.

In the late 1930s, emotions were running high over territorial rivalry between Campbell, Wilmott, and Heslop Harrison, and every new discovery of Heslop Harrison's was like a thorn in the flesh of the rival Hebridean botanists. In addition to the impression I was forming of Heslop Harrison as a man who made enemies easily, I began to wonder if Wilmott, as a man of influence in the botanical community of the 1930s and 1940s, was goaded into spreading doubt about the growing stream of new plants being discovered in the Hebrides.

Such doubt, however spread, was to last decades. In the early days of my research I visited a scientist who knew quite a lot at second hand about Heslop Harrison's activities from people who had been close to him. This man gave me a first hint of the power exerted over the botanical community by the fact that Heslop Harrison's surviving son and grandson were also botanists. He believed that they might in some indefinable way sabotage the career of anyone who attempted to criticise their forefather. At the time, I had had no contact with either the son or the grandson. So when this man, whom I'll call Dr O'Connor, said, 'Look, you'd better not quote me on some of this – Heslop Harrison's grandson is on a committee that could turn down a grant application from me,' I thought at first that he was joking. But he wasn't.

O'Connor was a typical example of the effect that the Heslop Harrison allegations had had on a generation of British botanists. There were also rumours, passed down from one academic to another in Heslop Harrison's department over the years since he had died. O'Connor told me how, when he was once considering taking up a post at Newcastle University, his colleagues had said, 'Don't go to Newcastle – they invent records there.'

Among the interesting information O'Connor gave me was the name of someone described as Heslop Harrison's protégé from the age of fifteen or so. His story was similar to John Morton's. This man lived in Birtley, and as a boy had been very interested in natural history. Over the years this interest had been nurtured and fed by Heslop Harrison to such an extent that the boy went to university and became a botanist and eventually a lecturer in Heslop Harrison's department. Here was someone I needed to meet, a man who had known Heslop Harrison for thirty years or so and would be in a good position to give me some insight into Heslop Harrison's character and personality, whether or not he could throw any direct light on the Rum affair. I wrote him a letter, in terms that unfortunately took it for granted that most people, including the addressee, knew of the accusations against Heslop Harrison, and I was astonished to get a reply that might have come out of the Heslop Harrison Angry Letter-writing Course, examples of which will come later.

Far from being able to throw any light on the Rum events, this botanist refused to believe that either Heslop Harrison, father or son, had ever done anything their mothers wouldn't be happy with. 'I can tell you,' he wrote, 'that during my long association with the Professors Heslop-Harrison* and in my professional career among botanists I have never known, or heard tell of, them acting dishonourably . . . It is outrageous that some malicious jokers have been pulling your leg.' It was my introduction to the surprising fact that several people who must have been quite close to Heslop Harrison not only did not believe that he was a villain – which was at least understandable – but claimed never to have heard any whisper of such allegations.

Another scientist, Alan Davison, who works today at Newcastle University, was both more forthcoming than O'Connor and more knowledgeable about the allegations against Heslop Harrison. He had been an undergraduate at Newcastle in the 1950s and met the professor, who had retired by then, when he came in occasionally while the undergraduates were identifying plants. 'He was bristly,' said Davison, 'and as a meek undergraduate I didn't take to him immediately.' But Davison had great respect for Heslop Harrison's scientific knowledge: 'His work on roses was amazing. We had all of these slides [of rose cells]. It was staggering how they worked out. They've got the weirdest chromosomes you could ever imagine.'

When Davison first met Heslop Harrison, he hadn't heard anything about the accusations of fraud. This was ten years after

---

* I had not in my letter to him suggested any wrongdoing by Heslop Harrison's son.

John Raven wrote his Rum report, and twenty years after rumours of Heslop Harrison's misdeeds had begun to spread through the botanical community. But later, when he became a member of the teaching staff, Davison's suspicions were alerted by two events. One concerned Heslop Harrison's son-in-law, Willie Clark, a botanist working in the department, who had married Heslop Harrison's daughter, Helena, known as Dollie.* Clark became a close colleague of Davison's, and Davison remembered him fondly. 'He was an old-style Scottish gentleman,' said Davison, 'the sort of person you very rarely meet these days. He would go out of his way to help people.'

Clark was to crop up at several points in the story as it unfolded, since he was an important member of the Hebridean expeditions and a co-author of many of the papers that reported the discoveries. One day Davison came across a reprint of a scientific paper with Clark's name on it as sole author, written in 1939, that was still lying around the department.[19] The paper had been published in a local botanical journal, *Vasculum*, edited by Heslop Harrison, and there was something odd about it that struck Davison and his colleagues. In its style and content it seemed unlike Clark. In fact, it had much more of the character of one of Heslop Harrison's papers – forceful, definite, brooking no disagreement.

The article described how Clark had submitted three papers on Hebridean plants to the editor of the *Journal of Botany*, and was surprised to find that they were handed 'to a referee who, to put it mildly indeed, was known not to be benevolently disposed towards our Hebridean researches' (clearly a reference to Wilmott).

The author of the article then described how he withdrew the papers, and he proceeded to tell the readers of the *Vasculum* what was in them. In the midst of some technical botanical details about a willow Clark had discovered, the writer made the following

---

* Confusingly, Heslop Harrison's son George, also a member of his department, married a woman called Dorothy, also known as Dollie.

peculiar reference: 'Strange as it may seem to the inventors of the "lost, broken or not collected" fairy tale, some of its twigs were carefully pressed and the rest of the specimen planted in the garden . . .' There was no reference anywhere else in the article to the phrase 'lost, broken or not collected,' which could only have meant something to a handful of readers and whose full significance I was only to realise much later in my researches, when it was associated inseparably with Heslop Harrison himself.[20] Toward the end of the paper, which described a bitter argument with Wilmott over the naming of Clark's willow, Clark/Heslop Harrison wrote: '. . . my record for *S. Harris* was rejected, not because Wilmott had seen it but because it was "outside" the known range of the plant . . . Such are the extreme views to which armchair botany leads!' An asterisk leads to a footnote, which says, 'But perhaps this new rule applies to certain people only!'

Everybody in the Newcastle department was convinced that Clark could not have written the paper. The abuse heaped on Wilmott and the liberal scattering of exclamation marks was just not Clark's style. Whereas it *was* Heslop Harrison's. Clark himself refused to be drawn on the matter. 'I did actually mention it to him jokingly at the time,' said Davison, 'and I just didn't get a response at all. And the odd thing was, there were a dozen or more copies of this paper around the department when I worked there twenty years after it was published. Willie must have known about it. Why did he allow them to lie around?'

The second event that really alerted Davison to the untold history behind Heslop Harrison and his activities occurred after Davison and two colleagues made a chance discovery while taking a group of students botanising on Teesdale: 'We used to go to the same places every year where all the interesting things were. Come this one year in the sixties we were up on Widdy-bank and the weather turned atrocious – the rain was horizontal. Willie [Clark] didn't want to give in and wanted to have lunch on the hoof. The students were getting very restless, so two colleagues, Oliver Gilbert and Tom Hutcheson, and I said, "Look, we've got to get some shelter," and we cut across an area we hadn't been in before. One of the students bent down and picked a plant and said, "What's this?" To which I said, "It's *Betula nana*,"* and walked on. I walked for several yards and suddenly thought, *Betula nana* doesn't grow on Teesdale . . . I called Tom and Oliver and said, "Look, this is *Betula nana*," and they did a bit of a wobbler. So we called Willie back, because he was marching ahead, and said we'd found *Betula nana*. We had a bit of difficulty finding it again; I think there was one bush. And he said, "But the nearest place is the north of Scotland!" It was an amazing record. Tom Hutcheson and I went back and dug into the peat under it and did some pollen analysis, and I actually found some fossils in the peat showing that it had been there for seven thousand years. So we thought that should be written up and Tom wrote it up.'

Such an unusual observation merited publication, but even in the 1960s it turned out that there was some resistance to accepting discoveries that came out of Heslop Harrison's old department, although he had long since retired. 'We started getting these rumours of people saying, "Hey, it's Newcastle at it again," ' said Davison. 'Tom was totally baffled by this and not a little annoyed, and I suddenly realised that there was all this feeling out there that had gone on in the past.'

So far, the only open criticism of Heslop Harrison's research I had read was in the Raven report. Apart from the rumours relayed many years later about the quality of his reports and observations, and the fact that a few of his plant attributions had been removed from the reference books after his death, I had seen no other indication that Heslop Harrison had behaved in any way other than that befitting a leading botanical scientist and fellow of the Royal

---

* Dwarf birch.

Society. Then, in the issue of the *Journal of Botany* for June 1941, I came across an article titled 'Studies of British Potamogetons' by J.E. Dandy and G. Taylor. Like so many of the official botanical names, the exotic-sounding 'potamogeton' turned out to conceal the more mundane pondweed. Dandy and Taylor had written a series of papers about the British pondweeds – this was the fifteenth, and there seemed no hint of a slowdown in their productivity. This particular article was dominated by complaints about the plant-gathering activities and slipshod reporting of Professor J.W. Heslop Harrison.

> It should be mentioned that *P. colouratus, X P. billupsii, X P. sparganifolius*, and *P. alpinus* have recently been recorded in a dupli-cated typescript sheet under the title: 'Occasional Notes from the Department of Botany, King's College, Newcastle upon Tyne, 2. No. 1. March 14th, 1941. Records of Hebridean Plants by J.W. Heslop Harrison, F.R.S.' This document is unpriced, and judged by its makeup can hardly be intended as a valid scientific publication. In case, however, this improvised sheet should be quoted as a published work we are obliged to point out that the parentage '*P. colouratus x P. natans*' attributed to *X P. sparganifolius* is an error; the real parentage is *P. gramineus x natans*.

It was only as I dug further into other botany journals of the period that I realised that this criticism was mild compared with some of the barbs directed by Heslop Harrison at his fellow bota-nists – in particular, at Wilmott and Campbell.

Heslop Harrison had the advantage of a periodical, the *Vasculum*, that seems from its content and flavour to have been his personal mouthpiece. Its title was a word unfamiliar to me. It was several months after starting my researches before I actually saw a vascu-lum. When I was talking to Max Walters at his house in Grantchester, I asked him if he had one, and he seemed surprised at the question.

'I've got several,' he said, and offered to show them to me. He led me out to a garden shed and pulled down from a shelf a hollow metal object like a flattened artillery shell case. It was rather battered and painted black, although the paint had come away in some areas and been replaced with rust. On the flattened side there was a door, hinged on a long edge and with a cumbersome catch made of stiff wire. The whole thing hung from a strap that you put over your shoulder, with the door on the outside opening out and down, and the idea was that when you saw a plant you wanted to take home, you swooped it up and popped it into the vasculum, closing the door to keep it in. And you needed to close the door: its design was such, it seemed to me, that if you put more than one or two plants in it, they could come above the rather narrow lip and fall out.

The first vasculum Walters showed me was about the size of an octavo book, the second twice that size, and the third big enough to contain an entire shrub. This was the one he had used when he was director of the Cambridge Botanic Garden.

These objects were clearly some kind of badge of office, the mark of a serious botanist. I can't imagine anyone turning up in the field sporting a brand-new vasculum, with shiny black paint, smoothly oiled catch, and not a speck of rust. I wondered if botanists, like schoolboys with a new cap or bag, would kick their new vasculum around a bit and rub it in the dirt, so that they didn't seem like neophytes among their colleagues. Because of its universal use in botany, the vasculum seemed an appropriate choice as the title for a botanical journal.

The *Vasculum* was where Heslop Harrison's homely accounts of tramping the botanical byways were published, and he also used it to launch attacks on other naturalists from time to time. Cynthia Longfield got the sharp end of Heslop Harrison's pen when she had the temerity to reply in a letter to the *Vasculum*[21] to a criticism he had made of a report she had written about dragonflies omitting several of his records from Durham and Northumberland. Her letter seemed to start rather contritely – 'I stand reproved before Professor Heslop Harrison' – but she went on to defend herself perfectly reasonably for the interpretation she had given of some of Heslop Harrison's published reports of dragonflies. Heslop Harrison, of course, had to have the last word. And what a word it was:

> When the above farrago of misstatements and errors was brought to my notice,* my first inclination was to leave it to speak for itself. However, as, shorn of the usual quaint and familiar catch-phrases, it forms a characteristic example of the attempts to bolster up impossible positions to which we have become so accustomed lately, it seems best to reveal its real hollowness . . .

After taking a couple of pages to rebut Miss Longfield's rebuttals, Heslop Harrison then finished with an astonishing final sentence for a paper in a scientific journal: 'Lastly, one would be greatly gratified to know the EXACT reason, or reasons, why Miss Longfield's letter was written.'

It seems that paranoia was at work here. Words like these are consistent with a belief of Heslop Harrison's that people were out to get him, people who would not stop short of using other naturalists, such as Miss Longfield, to do their work. And who could those people have been?

Well, we don't have to explore too many more back numbers of the *Vasculum* to find out.

In one *Vasculum* article, for example, Heslop Harrison took the opportunity to correct several recent papers about Hebridean

---

* This is not the only example of Heslop Harrison becoming aware of a critical comment only when it was 'brought to his notice'. We will notice the same thing happening with Raven's only published comment on Heslop Harrison's Rum 'discoveries'.

botany, some by Maybud Campbell. Of three Campbell papers, Heslop Harrison wrote:

> In some ways, the papers are misleading, and in no respect is this more evident than in the treatment of the commonest species . . . there are too many references like 'By Newton Lodge', 'Under wall outside the garden', 'Marsh below the garden', etc. . . . Here we wish to make it clear that no matter what Miss Campbell's predilections are for garden walls and fences, the authors of the paper she purports to criticise do not share them . . . Once again, we refuse to deal with garden fences . . . In connection with this species Miss Campbell has deliberately ignored Watson's warning that plants given in the Natural History of Barra, O.H. and not mentioned in paper (10) must be regarded as lacking in confirmation! . . . From these comments, it is clear that if the rest of 'the preliminary researches I have made for my Flora of the Outer Hebrides' (Miss Campbell's own words!) are as misleading as those set out in 'Notes on Outer Hebrides Records', then a foundation has been laid for profitless but necessary, corrective researches by some future worker.

However robust Campbell was in her own dealings with others, such scathing comments on her botanical writings can only have fuelled her dislike of Heslop Harrison.

In the same way, Heslop Harrison used the pages of the *Vasculum* to launch an attack on Wilmott. It arose from a paper of Heslop Harrison's, refereed by Wilmott, about some plants collected by the Newcastle group on the Hebridean island of South Uist, in particular an orchid that Heslop Harrison identified as *Orchis fuchsii*.

Heslop Harrison wrote:

> Whilst Wilmott, the 'referee,' most condescendingly admitted that there was very little to criticise in the paper, he objected to the inclusion of this species on the grounds that it was 'doubtful.'

As is well known, we have published a series of preliminary articles in preparation for our major Flora of the Inner and Outer Hebrides. Throughout these undertakings, I have been solely responsible for the Orchids, and this paper provided no exception. Knowing that in the case of this plant I was dealing with a special Hebridean form, I most carefully indicated that it was reserved for detailed consideration later. My surprise (or 'shock' as a friendly correspondent puts it!) may well be imagined when I state that the July number of the *Journal of Botany* contains an article from the pen of the so-called 'referee' discussing the plant to which objection had been taken! Without further comment that fact may be allowed to speak for itself.

Wilmott apparently thought the orchid might be a new species, and he gave it the name *Orchis hebridensis Wilmott*. We know enough about the situation already to be able to guess at the effect of Wilmott's 'discovering' a new plant that Heslop Harrison had thought was an existing one, and adding his name to it.

Heslop Harrison continued:

My acquaintance with the plant has been much more extensive than Wilmott's, for I have studied thousands in the field in Coll, Tiree, Rhum, Muck, Eigg, Barra, South Uist, Vatersay, Muldoanich, Fioray, Fuday, Flodday, Sandray and Pabbay. As the result of these intensive studies, I have no hesitation in stating that the plant cannot be regarded as specifically distinct from *Orchis fuchsii*, but, instead, must be considered as a well-marked Hebridean development of that species.

After more details in the same vein, the paper finished:

As Wilmott himself confesses in effect, the publication of the alleged species was quite premature. The whole episode simply confirms the logical view that (to use Dr Clark's happy phrase) armchair botany, spiced with a modicum of field work, is hopelessly inadequate in critical groups, and only serves to clutter up the literature with a series of untenable names.

The repetition of the description of Wilmott, one of the most senior figures in the British botanical establishment, as an 'armchair botanist' was, on its own, a casus belli that could not be ignored. As this was just one of a number of similar attacks on Wilmott and his supporters, it would only be a matter of time before some kind of response was organised.

5

The Plan

Volume LXXIX of the *Journal of Botany*, published in 1941, included two papers by Heslop Harrison and others that formed the trigger for attempts to unmask him. One paper was headed '*Carex bicolor all.*, A Sedge New to the British Isles, in the Isle of Rhum,'[*] with Heslop Harrison as sole author, and the other 'Further Observations on the Flora of the Isle of Rhum', by Heslop Harrison, R.B. Cooke, Helena Heslop Harrison, the professor's daughter, and her husband, W.A. ('Willie') Clark.

For Wilmott and Maybud Campbell, after four years punctuated by discoveries of plants new to the Hebrides or new to Scotland, this was the most dramatic discovery Heslop Harrison had ever announced, a plant never seen by them in their comprehensive Hebridean researches and, in fact, never before seen in the British Isles. They heard about the discovery some months before it was published and were clearly eager to see a sample. In April 1941, in

---

[*] One guidebook says that the original name was 'Rum' and that the Victorians changed it, adding an *h*, presumably out of the same censorious impulse that led them to cover up piano legs. Magnus Magnusson, in his book about the island, describes this name change as 'a spurious late Victorian whim.' Heslop Harrison and Raven both called it Rhum, but nowadays, in this more permissive era, it has reverted to Rum, the spelling I use in this book when I'm not quoting someone.

the Wilmott file, there is a letter from Heslop Harrison about the newly discovered sedge. (In spite of the 'armchair botanist' gibe, relations were on a reasonably even keel at this time.) The letter seems to be a response to a request from Wilmott for a sample. 'Of *Carex bicolor*,' Heslop Harrison wrote, 'we only took five specimens. One was sent to Kew (alive) directly from Rhum. R.B. Cooke has the second, and Clark the third. I brought two (alive) home, and one of these died after fruiting: it was given to my son and pressed. The other is a magnificent plant now and should produce many flowers. If so, as I promised Ramsbottom,* you shall have one good shoot.'

This shows that, whatever the origin of the first colony to be reported by Heslop Harrison, his garden now had a flourishing plant with many shoots, capable of being sent on to others or replanted in suitable soil. Sure enough, his next letter to Wilmott, written a week after the first, was accompanied by two seedlings of *Carex bicolor*. 'They are derived from the plant which died and have grown well,' Heslop Harrison wrote. 'The other of our original Rhum plants has grown like a cabbage!'

On the first page of his 1948 report to the Council of Trinity College, to whom he applied successfully for a grant of fifty pounds to make the trip to Rum, Raven summarised the growing amazement of the botanical world over the previous ten years at reports of Heslop Harrison's discoveries:

> The finding of these plants in the Hebrides, where none of them had ever been seen before, represents a very remarkable achievement, and has naturally added very greatly to Prof. Heslop Harrison's reputation. But unfortunately a number of the best qualified of British botanists (several of whom, incidentally,

---

* John Ramsbottom, keeper of the Department of Botany at the British Museum.

knew some at least of the Hebrides pretty intimately) came to view this ever-lengthening list of Hebridean rarities with growing suspicion. Since the Professor was reluctant to allow outsiders to examine some of his most notable specimens, the belief became widespread that some at least of his records rested on faulty identification; other records again were alleged to be due to confusion of specimens in the laboratory; and even the apparently fantastic rumour gained currency that the Professor must himself have planted the specimens that he subsequently discovered. There was, so far as I know, no decisive evidence one way or the other; suspicion and controversy were growing apace; and the time seemed to have come when, in the interests of scientific truth as well as peace, a determined effort should be made to settle the dispute.

How did the 'determined effort . . . to settle the dispute' come about? It's one thing for individual botanists gathered two or three together to exchange rumours or even strong suspicions, but Raven's application for funding to Trinity makes it look as if he was acting in some way on behalf of the whole botanical community. Could this have been the case? And if so, what triggered such a decisive move?

Did Wilmott have anything to do with it? It is difficult to believe that he didn't, since he had good reason to want something done. Even if there had been no personal antipathy, as Keeper of the British Herbarium Wilmott had to ensure that the records were correct and unpolluted by dubious attributions. William T. Stearn, who joined the British Museum botany department in 1952, described Wilmott as follows:

Wilmott was a botanical philosopher abounding in ideas for complicated, labour-intensive methods for doing everything and he became so devoted in the elaboration of method as to achieve little else; hence his relatively few publications, mostly in the *Journal of Botany*, do little justice to his fine logical intellect, his outstanding knowledge of the European flora and his industry in the herbarium and nothing, of course, to his prowess as a table-tennis player. The last explains how he came to possess such a quantity of table-tennis balls that the Museum bought them from him to construct models of molecular structure.[22]

We can learn something of Wilmott's temperament when Stearn compares him with one of his friends, Charles Carmichael Lacaita, in the following terms: 'Both Wilmott and Lacaita were robust characters with volatile temperaments, and contemporary accounts testify as to the vigour of their exchanges on botanical matters. It may be no coincidence that both of them met sudden ends through cardiac arrests.' In a pattern that recurs several times, Heslop Harrison seems to have picked as an opponent someone with a similarly mercurial temperament, and there is evidence in the Natural History Museum files that he had wound Wilmott up to a pitch of fury by the mid-1940s.

One factor in this is the continuing suspicion that Wilmott's position allowed him to discredit, suppress, or even pinch Heslop Harrison's discoveries, to his own glorification, as Heslop Harrison alleged in the matter of *Orchis fuchsii*. Without going into the minutiae of the issues, some quotes from a series of letters exchanged in 1945 give a flavour of the relationship.

Dear Wilmott,

You seem to have misunderstood my letter completely. You asked me some time ago for certain species. These I had as *seedlings* to spare, and these *seedlings* I offered you . . . it is perfectly natural that I should desire that no one should work at my material before my paper was published . . . As far as I know I never sent you a record of *Carex pedata*, but I did send one of *C.*

capitata. This I was clearly informed you had suppressed . . . I certainly object to you preparing a description which would anticipate my own work . . . A worker is entitled to work up his own discoveries . . .

In Wilmott's reply, he said that he could not think why Heslop Harrison thought he wanted to 'crib' and complained about the difficulties of trying to do his job in the face of abusive and intemperate criticism. This opened the floodgates. Heslop Harrison wrote:

Let me draw your attention to a few facts.

1. No abuse was contained in my letter. I took up points in which you abused me and answered them.
2. In the terms of abuse you applied to me, 'intemperate criticism', 'make it very difficult', etc., etc., I recognise phrases which, with many others, have been used to describe you and your acts – and this from many parts of the country!
3. You write of 'usual co-operative manner'. When has there been any attempt at such co-operation on your part? We have never at any time received help from you. My recollections are of your failure to acknowledge specimens, of threats to ignore our work (letter in my possession!), intemperate criticism of our work and of other attempts to discredit us.
4. I quite realise the import of the silly placing of brackets around records,* the attempts to establish the position of super-referee, the curious rumours of which the sources are known, etc., etc. In spite of these vagaries, I have sent you specimens, have kept others for you, have supplied

---

* The placing of square brackets around reports of botanical discoveries signifies that they have not been corroborated by anyone else. Heslop Harrison accused Wilmott of doing this to his records without justification.

information when asked to do so and so on – all because I
wished to show goodwill . . .

. . .

7. I disclaim any responsibility for unpleasantness or abuse. Like
every botanist who raised the matter with me I see clearly
that our sins lie in the fact that we commenced a research [*sic*]
and have persisted in it with success.

In his reply, Wilmott attempted to stem the torrent of abuse but
failed. 'Dear Harrison,' he wrote, 'I could quote abuse from your
letters but do not intend to deal with other than scientific
matters . . .' He then did so, and finished the letter:

You may guess as many wrong answers as you like, but what I
now say is the simple fact. It* was not suppressed. You are now
suppressing it.

The other matters can be left. I am not officially concerned.
What you say about super-referee is just nonsense. I only want
to see material and feel free to express my views. If you force me
to use [. . .] I must continue to do so.

Yours sincerely

J. Wilmott

The short letter from Heslop Harrison that closed the correspondence ends:

One very important British botanist stated in connexion with
your bracketing efforts: 'The bracketing of the record reflects
more on the mentality of the bracketer than on the record.'
You would be somewhat astounded if you knew who said
that.

---

* The record of *C. capitata*.

It is sincerely to be hoped that the insertion of the material concerning *Carex pedata* gives you that glowing sense of importance for which you yearn. I cannot think of anyone else who would revel in it as you do. Again I had been informed of your adopting a similar attitude with very different botanical affairs.

Yours sincerely

J.W. Heslop Harrison

Across the top of this last letter, in Wilmott's handwriting, is scribbled the one word 'IGNORED!'

To put it no more strongly than this: the animosity expressed in the correspondence would have made Wilmott only too happy to assist in any effort to discredit Heslop Harrison and to mobilise other botanists to help achieve the same end.

There is no indication of when such a plan was first laid. One possibility is that the idea arose on the Glen Affric botanical excursion in 1947. This was a significant event in British botany that was planned in the months following the end of the Second World War, as life began to get back to normal and people were able to think about resuming activities that, in wartime, might have seemed too trivial or unimportant. The particular stimulus for the excursion was a plan to tap hydroelectric power from a system of rivers that ran off the northwest Highlands of Scotland. The construction of two dams and two tunnels would disrupt much of the plant life in a large area around the glen and submerge some of the valley sides under the reservoir that would build up behind the dams. The concern of British botanists at the possible effects of this scheme on Glen Affric was one of the earliest signs of a post-war interest in conservation issues, which was to grow into the environmental movement.

An expedition was organised to survey the glen before construction work began, and some of Britain's leading botanists were among the group. Charles Raven was one of them, accompanied by his son, John, then aged thirty-two. Also on the expedition was a famous botanical trio, who later wrote the standard guide to British flora, known as Clapham, Tutin, and Warburg. A.R. Clapham and T.G. Tutin were professors of botany at Sheffield and Leicester, and E.F. Warburg was at Oxford. With all the participants dead, we can only speculate about a possible origin of John Raven's trip to Rum during late-night discussions over warming cups of cocoa, as damp socks were unfurled and feet dried in front of a log fire in some spartan Scottish hostel.

In the Wilmott file in the British Museum archives, there is a letter from Canon Raven to Wilmott written in 1947, after the Glen Affric expedition, indicating a long and close friendship between them: 'I enjoyed our days in Scotland quite immensely,' Canon Raven wrote. 'You are the only person in the world who shares my remembrance of climbing lamp-posts in Cambridge in the early years of the century . . .'

With such a friendship between Canon Raven and Wilmott, and a shared passion for botany, it is not difficult to see how the canon's son might have been roped in to deal with the increasingly reprehensible – as they saw it – deeds of Heslop Harrison.

There is some indication in a letter written to John Raven after the Rum trip that Wilmott had first broached the topic 'after the war'. In all probability it was at Glen Affric that detailed plans were laid for the trip that actually took place, involving several of the botanists who had been on the expedition. That Raven was to be one of the party is an indication of how highly his botanical skills were valued, even though he was always ready to point out in his writings that he was an amateur.

The task of helping to unmask Heslop Harrison would have attracted Raven on two levels: his spirit of adventure would have been aroused at the possibility of such a 'lark'; and the moral worth of the cause would also have appealed to him. He loved plants, he painted them, and he fulminated against collectors who tore them out of the ground; the possibility that someone had faked botanical data would have provoked his anger. But he didn't allow his emotions to show in the way he wrote his report. His description of events is calm and reasoned.

Raven showed no signs of the sort of fierce ambition that drove Heslop Harrison. Sometimes envy of another's successes can lead someone to irrational actions, but Raven's career was as a classicist, and a very good one. Indeed, even if Raven had professional botanical ambitions, Heslop Harrison was no threat. And Raven's accounts of the work of other botanists suggest that he was capable of praising people fully when they deserved it.

Raven's task was to inspect as many as possible of the sites and plants that Heslop Harrison had claimed as 'discoveries' on Rum. He listed these plants at the beginning of his report to Trinity College:

> For the past 13 years a party of botanists and students, under the leadership of Professor J.W. Heslop Harrison, D.Sc., F.R.S., has gone out each summer from King's College, Newcastle-upon-Tyne, to investigate the flora of the Hebrides. During this long period they have published, at regular intervals, and in a variety of periodicals, a series of very remarkable botanical discoveries, in both the Inner and the Outer Hebrides. These discoveries have consisted, in the main, of two separate elements: first Arctic-alpine species, at least 5 of which were new to Britain, while many others were known only from single localities on the Scottish mainland; and second, Southern species, known in the British Isles, if at all, only in the South-Western corner of England and the Channel Isles.

> The former of these two elements contains, for instance: –

> *Epilobium lactiflorum*
> *Erigeron uniflorus*

Carex capitata
Carex bicolor
Carex glacialis

All new to Britain, and all recorded from Barkeval, Isle of Rhum.

Lychnis alpina – known only in 2 or 3 very small areas on the mainland
Carex microglochin – known only in Coire Buidheag, Ben Lawers
Carex lachenalii – known only on the highest Scottish mountains
Carex chordorrhiza – known only at Altnaharra, in Sutherland

The latter element contains only 3 plants that are relevant to this report, namely: –

Juncus capitatus – known only in the Channel Islands, West Cornwall and Anglesey
Polycarpon tetraphyllum – known only in the Channel Islands, Cornwall, Devon and Dorset.
Cicendia pusilla – known only in the Channel Islands.

This, then, was the rogues' gallery – the roll call of plants that, over the years, had created a growing sense of astonishment in the botanical community at Heslop Harrison's amazing luck – or skill – in coming across such rarities at regular intervals. They included a willow herb, a fleabane, several sedges, a couple of types of pink, and a rush, and none of them had been seen before in the vicinity of the sites where Heslop Harrison claimed to have discovered them.

In fact, even before Glen Affric, Raven had tried to find some of Heslop Harrison's 'discoveries'. In 1946, on the Isle of Raasay, just off Skye, he spent several hours searching 'a small stony patch' on which Heslop Harrison had claimed in print to have found two of the plants on Raven's list of 'very remarkable botanical discoveries' – *Juncus capitatus* (a tiny rush) and *Cicendia pusilla* (a small member of the gentian family). 'Not only did I fail to find a trace of either plant,' Raven wrote, 'but I left the spot convinced that they were not there. It is true that in his published account the Professor had himself reported that 'almost immediately' after the discovery of the plants 'the bulk was badly damaged by cattle'; but since both are annuals I did not believe that they would be so easily and suddenly exterminated.'

As an excuse, 'the cattle destroyed my plant' seems to rank alongside 'the dog ate my homework', and Raven's suspicions were reinforced by the fact that, as annuals, the plants would have had a strategy for putting seed into the ground, so that even if an individual plant was destroyed, its seed would have produced more plants the following year.

There was one other odd instance that fed Raven's growing suspicions. A friend of his, Professor J.R.M. Butler, had visited Rum in 1947 and met Heslop Harrison, who took Butler a short distance up the Kinloch River and showed him a small plot of land in which he was cultivating some of the island's rarest plants. Butler wrote the names of the plants on the back of an envelope. One of them was the increasingly familiar *Juncus capitatus*. Of course, there may have been an innocuous explanation for this garden of rarities – Butler implied that Heslop Harrison had said it was for 'purposes of study' – but to an already suspicious mind, this odd event provided further fuel. (It has to be said, however, that if Heslop Harrison was growing the plants to *re*plant as newly discovered rarities, he might not have been so open with a visitor. But then, by this stage, more than ten years after the wave of Hebridean discoveries had started, perhaps he felt immune to discovery.)

Incidents such as the above turned the murmurs of suspicion into a roar. An expedition had to be mounted to try to discover the truth. Along with John Raven, there would be Maybud Campbell, Clapham, Tutin, and Warburg. But by a process of human evaporation, these pillars of professional botany dropped out one by one from the plan. Whether they had realised what a hot potato their trip would become or were merely victims of chance, the party of five became three when Tutin had to take some of his students to Switzerland and Warburg suddenly chose to get married.

The three remaining members of the party – Campbell, Clapham, and Raven – decided to go first to Kilchoan, an area of mainland Scotland similar to Rum but botanically unexplored. They hoped either to find examples of plants that were the same as the ones Heslop Harrison reported on Rum or, purely for their own botanical interest, to find other new plants. They would then go on to the Isle of Harris, from which three surprising sedges had been reported by Heslop Harrison, to see if they were still there. Finally, as a climax to the tour, the three planned to stay on a cabin cruiser moored off Rum, from which they could conduct further investigations.

The next setback to their plans was that the cabin cruiser became unavailable, and it proved impossible to find a substitute. This might not have seemed a mortal blow if Rum hadn't been privately owned, by a Lady Bullough. There was no commercially available accommodation, and Heslop Harrison had effectively sewn up the situation, it seemed, as far as camping or hiring a cottage was concerned. Since the party did not want to create a fuss by their presence, given the nature of their task, there didn't seem any way they could stay on the island itself.

At this point Raven was about to give up and hand the fifty pounds back to Trinity. Then another idea occurred to him.

It so happened that Charles Raven had been pursuing his own botanical interests in a correspondence with Heslop Harrison on matters unconnected with John Raven's quest. During this correspondence, he mentioned that his son was hoping to visit Rum during the summer, and Heslop Harrison, unsuspecting, replied – 'cordially enough', John Raven says.

Certainly, Heslop Harrison's letter to Charles Raven is a model of goodwill:

> I shall be very pleased to help your son when he is on Rhum. If his visit coincides with mine it will be delightful to take him to the stations for all of the rarities. If he is there when I am else-where I am perfectly willing to give him full directions if he gives me a clear assurance that the maps supplied will not under any circumstance be used directly or indirectly by any other individual than himself. The quantities of the plants involved in some cases are extremely small and I am strongly averse from being a guide to over-collecting and extinction.

All pretty reasonable stuff.

Two weeks later, Heslop Harrison wrote directly to John Raven. He is still 'very pleased' to assist Raven in every possible way, but a warning note creeps in at this stage:

> The greatest snag on Rhum is that there is neither accommodation nor food available. As far as I am concerned I have rooms at the Castle whilst my party rough it in a small house. Unfortunately, as is absolutely essential, the personnel of the party is already arranged and the available space fully taken up. The fact that there are three ladies with us is an awkward factor. However, if you can get a tent and sleep in that we could possibly deal with one more person feeding with us. <u>You would however have to obtain permission to land</u>; otherwise you would find it impossible to come ashore.

When Raven replied to this letter, he expressed some surprise at the limited accommodation and food possibilities on the island, leading the professor to make the point again more emphatically:

You should realise that there are only about six inhabited houses on Rhum and that, in August, all extra accommodation in them, which is excessively small, is taken up by relatives – sons and daughters who have left to work elsewhere. Furthermore, there is no food to be had anywhere: everyone on the island has to arrange individually for his household's supply to be brought by boat. We send a lot of food in advance and in addition take emergency ration cards for a fortnight to Maclean, West Highland Stores, Mallaig. All food obtained on these cards we carry with us. You will have to do the same, and you must make sure you have enough food to make you independent of anyone else's supplies.

Clearly Raven was not going to be able to borrow as much as a cup of sugar from the professor's party. And in case he still hadn't got the message about accommodation, Heslop Harrison gave it once more: 'You will have to take a tent as I have no spare accommodation. Mrs Heslop Harrison and I sleep at the castle. Two ladies use one of my rooms in our cottage and three men the other – all sleeping on the floor. We feed together in a single room downstairs.' There was no longer any mention of the possibility of Raven eating with the others. In fact, in a P.S. that sounds as if he was addressing someone he thought feeble-minded, Heslop Harrison wrote, 'Please note: 1) Knife, fork, spoons, plate, cup and small bowl will be needed. 2) All food should be sent well in advance as parcels to Rhum take a longish time for the journey. 3) I have cooking utensils but a spare Primus is useful!'

The final letter in this first phase of correspondence, written from Rum as Heslop Harrison set off around the Hebrides, finds the professor in a more mellow mood again: 'We arrive here on Monday August 3rd and will see you then. As I told you, you can feed with us as I know how awful camping alone and preparing food after a hard day's work can be. We have discussed a camping site for you and have fixed upon one just behind our house.' Then, with a message that is beginning to sound monotonous, Heslop Harrison added: 'Don't forget to bring plenty of food (as there is none here!), a map – and dried milk.'

I have described this correspondence in detail because in later events it became an important issue. It is clear that Heslop Harrison was taken in by Raven's request to visit and had no suspicion that other people might be involved. When all became clear later, he was to nurse a permanent grievance against Raven for deceiving him as to the purpose of his visit; Raven referred often to his own feelings of guilt that he had only got onto the island because of the kindness of the man he was setting out to destroy.

But set out he did, on 12 July 1948, for a journey on which, for the first time, the paths of the professor and the don would cross.

6

The Forbidden Island

Rum is one of the Inner Hebrides, a group of islands, some very large, some small, scattered along the northwest coast of Scotland. They include Skye, Eigg, and Muck, as well as Rum. Beyond them, separated by a twenty-to-forty-mile channel, are the Outer Hebrides, which stretch from the large Isle of Lewis in the north to the tiny island of Barra in the south. Today Rum is a nature reserve, run by Scottish Natural Heritage, and receives visitors by the small boatload from Mallaig, on the Scottish mainland, some of whom spend a couple of hours, some a night or two, at Kinloch Castle. Most of the castle accommodations are less grand than you might expect, since they're in what were the servants' quarters, but a few of the grander bedrooms are available for a higher rate. There are no roads to speak of on the island and few cars. Only the Nature Conservancy Land Rovers occasionally bump over the stony tracks, and it's made very clear to the visitor that, apart from the occasional trip from the jetty to the castle with luggage, nothing short of a broken leg will get you a ride for any distance. You are there to walk, up hill and down dale, and the warning notices around the castle make it clear, in only a slightly less admonitory tone than Heslop Harrison's advice to Raven, that you should not expect more than the basic amenities during your time on the island. Having said that, the incongruously named Bistro near the castle kitchen supplies pleasant three-course dinners and wholesome packed lunches for those visitors who last longer than a day.

To a non-botanist arriving on Rum, the whole botanical enterprise seems a marvel. It defies understanding that anyone can painstakingly elucidate the differences between a huge variety of plants, finding somewhere in the forty square miles of territory a clump or even a handful of plants new to botany. To the untutored eye there are three types of small plant on Rum – green, purple, and multi-coloured. The green is grass, the purple is heather, and the multi-coloured plants are small flowers. Going beyond this to differentiate between the 2,200 species of plants requires an expertise that both Heslop Harrison and Raven had to a high degree. And even plants that might be grouped by some of us under the heading 'grass' – plants that are at the centre of the Heslop Harrison story – turn out to have unsuspected subtlety.

If you open *Grasses, Sedges, Rushes and Ferns*,[23] a fairly standard handbook for people interested in natural history, this is the sort of thing you will read:

> Grasses, it has to be admitted, have extremely complex flowers . . .
>
> Rushes have the simplest flowers of the plants in this book. They may look obscure and difficult but close examination shows that . . .
>
> The structure of sedge flowers is less obvious to the untrained eye than that of rushes. They have no recognisable petals or perianth . . .

Several of the plants that Heslop Harrison had claimed to discover as new to Britain were rushes and sedges. Any suitable area of land in the British Isles might have a few species of rushes and up to a score or so of sedges, but each area would be characterised by a slightly different collection of species. The Outer and Inner Hebrides were no exception, and Rum, as an island that had been in private hands for a hundred years or more, was a particularly unbotanised piece of island territory and provided scope for new findings. It also, for as long as it continued in private hands, provided limited opportunities for independent confirmation of any findings.

Because the British mainland flora had been surveyed and classified by the early years of the twentieth century, it was the turn of the more outlying parts of the country to provide scope for those botanists who liked to find new plants, or old plants in new places. In addition to remoteness, the Hebrides had the attraction of a uniquely distinct flora and fauna, the result of all the changes in topography that took place during the last ice age.

For all these reasons, the Hebrides provided Heslop Harrison with fine teaching material for his students, many of whom became the next generation of distinguished and passionate botanists. In the mid-1930s, he started regular expeditions to the isles, which led to a series of papers published at intervals of a few months over the next twenty years.

Unfortunately, it seems that just about the time Heslop Harrison's eyes were turning toward the Hebrides, so were those of the Ferdinand and Isabella of botany. Heslop Harrison began to take expeditions to the Outer Hebrides in 1935, and it was in 1936 that Maybud Campbell published an account of her first visit to the islands, to be followed a year or so later by her collaborator, Wilmott. As Andrew Currie, a modern botanist with the Nature Conservancy Council on Skye, puts it:

It would be futile to speculate what might have been achieved had these two separate teams felt able to cooperate. That there was a degree of enmity can be illustrated by reference to short papers by J.W. Heslop Harrison *et al.* (1938) and Campbell (1939b). A clash of personalities must have been a factor, one feels.[24]

Campbell's paper referred to by Currie, and published in the *Journal of Botany* in 1939, expresses its degree of enmity in a terse paragraph:

> In the last issue of this Journal, Prof. J.W. Heslop-Harrison, Miss Helena Heslop-Harrison, Mr R.B. Cooke, and Dr. W.A. Clarke [*sic*][25] state that 'it is proposed to place on record the plants . . . which appear to be absent from published lists for v.c. 110'.[26] The preliminary researches that I have made for my 'Flora of the Outer Hebrides' show that a number of species so recorded had already been published. They are as follows . . .

She then itemises fourteen of the Heslop Harrison team's 'unpublished' plants, showing that they had all been published, in references that go back as far as 1830.

Eyeing each other's discoveries and publications, the two groups worked away during the late 1930s at their flora of the islands, although there was a major disparity in their published output, with Heslop Harrison and his team firing off papers in all directions, while Campbell just about managed to give birth to *The Flora of Uig*, describing in some detail the vegetation of one area of the largest Outer Hebridean island, Lewis. Since each group knew of the other's activities, it's likely that the 'personal enmity' mentioned by Currie introduced a spirit of unhealthy competition as one party tried to out-botanise the other.

Although Heslop Harrison's group roamed widely throughout the Inner and Outer Hebrides, there must have been a particular attraction about Rum. Today, it has been described by Morton Boyd as 'one of the best documented islands in Britain', with detailed published surveys of its vegetation, insects, birds, and mammals, and its geology. But in the 1930s, Heslop Harrison was one of the few biologists to venture onto the island, and he jealously guarded access once he was in de facto control of its natural history. In fact, it wasn't a particularly welcoming place for the casual visitor. There was no public accommodation, and, as Raven discovered, if you got as far as the bay near Kinloch Castle, you still needed the owner's launch to make the final landfall. As a result of the Clearances, the eighteenth- and nineteenth-century 'ethnic cleansing' of areas of Scotland by landowners, the island has been sparsely populated for 150 years or more.

In fact, of the 119 named islands in the Outer Hebrides, only sixteen are permanently inhabited, with a population of crofters and fisherman, although nowadays there are also pockets of military and oil industry activity. Rum had a tiny population, down to twenty or so in 1936 from a high of 445 at the end of the eighteenth century. When John Bullough, a self-made Lancashire industrialist, fell in love with the island (but not its inhabitants) in the 1880s and bought it, he retained only three of the island families and imported farmers from the Scottish lowlands whom he saw as 'men of energy, intelligence and capital' compared with the 'well-fed, pampered crofter [who] continues to loaf away his time while his wife does the work'. Bullough would sit on the hillside and, while enjoying an after-lunch cigar, compose doggerel expressing his love for the island. One verse is more than enough to give the flavour:

> *And year by year as round we come*
> *To greet our grand old father Rum,*
> *He'll o'er and o'er renew his blessing*
> *Well pleased to see us to him pressing.*

John Bullough died in 1891 in a London hotel and left Rum to his younger son, George, along with £500,000, the equivalent of £50 million in today's money. It was George whose profligacy was to transform Rum from just another Scottish island with natural flora and fauna largely undisturbed by the activities of man into a most unnatural playground for himself and his rich friends for a few weeks in the year. George Bullough built an outrageously opulent house that would not have been out of place in a big city like Edinburgh or London, with visitors, staff and facilities whose necessities and luxuries had to be transported with difficulty from faraway places on the mainland. No expense was spared and no innovation went unexploited, including the first internal telephone system in a private residence in Scotland, with a ten-line exchange and the phone number Rhum 1, clearly of symbolic significance alone since there was no connection to the outside world.[27]

And it wasn't only the accommodation that seemed unnatural on this rocky island. Nature itself was transformed, as the Bullough family tried, and sometimes failed, to introduce new and exotic forms of animal and vegetable life for the amusement or sport of invited guests. The plants that Heslop Harrison introduced to the island, perhaps from his garden in Birtley, were insignificant and barely perceptible additions to the biology of the island compared with terrapins and bananas, deer and trout, alligators and humming-birds. Many of these were confined to glasshouses or ponds, but Bullough's behaviour – and the sporadic nature of his visits – still suggests that Rum's natural capacity to enthral or amuse needed some shoring up.

When George visited Rum for the first time as its owner, after the death of his father in 1891, he decided that he needed a much more palatial residence than the existing one, called Kinloch House, where his father had stayed and which was now rather dilapidated. With more money than sense, he embarked upon the construction of Kinloch Castle, which was completed in 1900.

Today the castle is opened sporadically to the trickle of visitors to the island, and when I visited, the castle manager, Clive Hollingworth, delivered a polished guided tour, with facts and figures at his fingertips. The estimated cost of the castle, £250,000, is the equivalent of £25 million in today's money.[28] Bullough thought nothing of importing red sandstone and soil from the

Scottish mainland and workmen from Lancashire to build the house and establish the garden. (When Hollingworth said that the red stone was imported from Ayrshire, I overheard a visiting American couple express their amazement that the stone was imported from *Asia*. Even if it had been, it wouldn't have made much of a dent in Bullough's budget.)

Hollingworth's tour was designed to shock visitors with its hints of Edwardian debauchery. Purveying a great deal of plausible but often unverifiable detail, Hollingworth described how George imported flocks of partygoers of both sexes by steam yacht in the summer, sometimes including King Edward VII himself. The highlight of the guided tour was the unfolding of the story of George's marriage to Monica Charrington, née de la Pasture, a beautiful French divorcée whose delightful portrait, clothed, hangs over the fireplace, while an unclothed one is more discreetly located in the private apartments. In spite of a rumoured attraction to his stepmother, his father's much younger second wife, one story goes, George was believed to be homosexual, and therefore people were very surprised when he married a young woman who could have had her pick of the nation's young men. The proffered explanation is that Monica had been one of Edward VII's many mistresses and that George agreed to be named in her divorce suit and then marry her, to save embarrassing the King, who was really the adulterer. Magnus Magnusson, in his book *Rum: Nature's Island*,[29] investigated this exciting story and failed to find a scrap of evidence for it, but it certainly spices up the tour of the castle.

As you walk along the echoing corridors and bump into the faded, worn, overstuffed furniture, you could believe every detail of Hollingworth's genteel but raunchy account of the doings of the Bulloughs. The bedroom corridor hints at night-time wanderings by specially invited male guests in the general direction of Lady Monica's bedroom, conveniently far from the suite of private apartments occupied by Sir George. And as the small knot of visitors was shown around the ballroom downstairs, Hollingworth pointed out the windows, placed with their sills above eye level so that the 'well-fed, pampered crofters' should not see the frightful goings-on inside.

After all this recreation of turn-of-the-century Rum partying, it's a bit of a let-down that the impressive steam organ, or Orchestrion, installed by Sir George in 1906, plays a rather staid selection of music. As George and his male and female friends took their after-dinner coffee in the galleried hall, they had to make do with such innocuous songs as 'The Honeysuckle and the Bee' and 'Home, Sweet Home.'

By the time Heslop Harrison first arrived on the island in the 1930s, Bullough seems to have settled with Lady Monica into a less flamboyant lifestyle – perhaps even a conjugal one. Visits to Rum by the Bulloughs were less and less frequent, and parties were few and far between. Staff left and the house and gardens became neglected. Sir George Bullough died in 1939, but his widow was to live until 1967. Throughout the 1940s and 1950s, she still spent some of the summer at Kinloch Castle, where she graciously granted Heslop Harrison permission to come to the island with his students and colleagues in the interests of botanical science. Heslop Harrison in his letters to Raven shows the kind of awe of Lady Bullough that might have been perfectly justified in the light of her ownership of the island but might also have been exaggerated deference from the son of an ironworker to someone with a title – a title, paradoxically, that owed its origins to money made as a result of the Industrial Revolution.

While the rest of the Hebrides were open to everyone who wished to explore the islands, including Wilmott and Campbell, Rum was off limits to anyone who didn't have the permission of the Bulloughs, and Heslop Harrison's correspondence with Raven suggests that he guarded his privileges jealously. It suited Heslop Harrison – it would suit any botanist – to have de facto control of who visited the island through his relationship with the Bulloughs. But a good scientific reason could also be offered in support of his jealous protection of territory. Both to safeguard his own scientific research and, more generally, to avoid harm coming to the rarer species, it was important that Rum not be overrun by strangers, perhaps even unscrupulous botanists who might pick plants for their own collections instead of leaving them in their natural habitats.

And when it came to enlisting Lady Bullough in protecting 'his' territory, since she was neither a botanist nor a distinguished professor, it is likely that she would allow him to take what measures he wished. There exists an interesting letter to Raven on this point from Lady Bullough's nearest neighbour, Dr John Lorne Campbell, a keen entomologist and owner of the island of Canna, a mile or so to the northeast of Rum. Raven consulted Campbell about the logistics of getting onto Rum and received a very supportive reply:

I am strongly of the opinion that further independent botanising and entomologising should be done on Rum. If you wish to do this overtly, it will be necessary to write to the proprietrix, Lady Bullough, or to the factor, Mr D. McNaughton, Isle of Rum. There is no certainty that your request will be granted; I am under the impression that the party who usually goes there* considers it something of a closed field, or monopoly, and outsiders may not have much chance.

Then, entering into the spirit of the chase, Campbell said:

If you wish to enter by the back door, this has been done, by campers who landed on the isle of Canna and then hired the boat of Mr Allan MacIsaac crofter to ferry them to Rum, where

---

* That is, Heslop Harrison.

they remained unbeknown for several days and were ferried back again, getting the mailboat the next morning. I suggest you write to Mr MacIsaac direct about this.

Campbell, a talented amateur entomologist, also added his name to the growing list of people who were suspicious of some of Heslop Harrison's discoveries, entomological as well as botanical. Heslop Harrison had reported finding a butterfly known as the Large Blue on Rum. As with many of the plants, this was a surprising finding, since no one had ever seen this particular butterfly anywhere in Scotland. 'I have heard the story of the Large Blue on Rum but remain sceptical,' Campbell wrote to Raven, 'not that substantiation of the claim would not be most interesting. My theory remains that the specimens were confused in the laboratory.' Twenty-five years later Campbell was to carry out his own investigation and, unlike Raven, publish the results.

Although Rum was to be the focus of Raven's inquiries, he and his colleagues approached the island at a leisurely pace. Heslop Harrison and his party were due on the island on 2 August, but in mid-July 1948, Raven, Maybud Campbell, and Arthur Clapham convened as planned at Kilchoan. They found three surprising plants, but since none of them figured in Heslop Harrison's *Flora of Rum*, these discoveries threw no light on the problem being investigated. Then Raven's group, accompanied by three botanical students, travelled to the Isle of Harris, where Heslop Harrison had reported several new findings.

As we will see, the minutely detailed account of Raven's investigations set out in his report provides very clear reasons for believing that many of Heslop Harrison's reported discoveries were not what they seemed. But that is with the benefit of hindsight. Before the trip, how could Raven have been confident that he would find Heslop Harrison out? As a non-botanist, I could not imagine where I would start looking on a forty-square-mile island if I suspected that a few of its plants were not what they seemed.

Raven of course had two advantages that I don't: trained senses, particularly the sense of sight, that were tuned to the clues provided by roots, leaves, and flowers, as well as rocks and soil; and a memory that had practised for two decades what psychologists refer to as pattern recognition, whereby the regular occurrence of several different features is immediately recognised as significant in a way that any individual feature might not be. It's rather similar to the way the police profile a criminal by different aspects of his modus operandi and ascribe a series of unsolved crimes to one perpetrator. Similarly, the natural occurrence of a plant in a specific location is usually accompanied by a 'pattern' of other facts – soil type, neighbouring species, terrain, temperature range, latitude, and so on. The more pieces of the pattern that are found, the less likely it is that the plant was introduced from outside.

An example of the way Raven worked is provided by his account of the trip to the Isle of Harris in the Outer Hebrides from 19 July to 30 July. His self-imposed task was to search for three rare sedges that Heslop Harrison had reported on the island. At the risk of exaggerating the difficulties faced by Raven – and thereby magnifying his achievements – I have to tell you that sedges are not the most flamboyant of plants. In fact, the seventy-three sedges described and illustrated in *Sedges of the British Isles* by Jermy, Chater, and David[30] all look very much like stalks of grass with little blobby clusters of seeds at the top. That is because they *are* all stalks of grass with little blobby clusters of seeds on top.

Heslop Harrison had reported finding three rare species of sedges on Harris – *Carex microglochin*, *Carex lachenalii*, and *Carex chordorrhiza* (the Latin names of sedges all begin with the genus name *Carex*). Raven was familiar with these species in their previously known habitats on the mainland, and he was able to use his knowledge of the environments in which they grew to establish to his satisfaction that not only could he not find them on Harris, they were also extremely unlikely ever to have grown there. He did this by looking for the plants' patterns – such characteristics of the landscape as the soil, the altitude, and the other plants that grew there. Specific plants grow best in soils of a certain acidity or alkalinity, prefer the temperature variations at certain altitudes, and are often accompanied by a characteristic group of other plants with similar tastes. So knowledge gained by studying the landscape and being familiar with the climate and soil can reinforce the likelihood of finding a particular plant, or emphasise the improbability of doing so, before any inspection, usually on hands and knees and sometimes with a magnifying glass, is carried out.

Here's how it worked with the very first of Heslop Harrison's 'discoveries' that Raven looked for:

I have little hesitation in saying that 2 at least of the 3 sedges are not where they are alleged to be. I cannot of course claim that I examined every inch of the areas in question; and even if I had I would never feel confident that I had not overlooked a plant so small and inconspicuous as *Carex microglochin*. *Carex microglochin* grows on the Ben Lawers range* in base-rich, micaceous bogs, accompanied by varying quantities of such other Arctic-alpine species as *Juncus triglumis, Carex saxatilis* and *C.† ustulata*, and *Kobresia*. Uisgnaval Mhor, the mountain in Harris from which it was recorded, is a dry and intensely acid‡ mountain. Its alpine flora consists only of *Saxifraga stellaris* and *Epilobium alpinum, Polygonum viviparum* and *Oxyria digyna, Salix herbacea, Luzula spicata* and *Carex rigida* and a very small quantity *of Arabis petraea*. This is a wholly different plant community from that which

---

* On mainland Scotland.

† That is, *Carex*.

‡ The opposite of 'base-rich' in the previous sentence.

accompanies *Carex microglochin*. *Juncus triglumis* and *Carex saxatilis* would, I am quite sure, be found always to accompany *C. microglochin* wherever else it were to be discovered in Britain. Neither of the two has ever been recorded from Harris, not even by Professor Heslop Harrison himself. Conversely *Arabis petraea* has never been found on the Ben Lawers range. Yet it is found, as Professor Heslop Harrison admits, on Uisgnaval Mhor.

I have quoted this somewhat technical paragraph in full because it shows the careful way in which Raven builds up his case for the nonexistence of *Carex microglochin* on Harris. He is relying on the sort of arguments we might use if a polar bear had been reported in the Sahara. Armed with a knowledge of the preferred climate, diet, and habitat of polar bears, you don't need to have scoured the entire nine million square kilometres of desert to be fairly sure that there are no polar bears there, however impeccable the credentials of the person who claimed to see one. And even if the discoverer had captured the bear and shown it to you in a cage in his London garden, it wouldn't really strengthen your belief in his story. In fact, as we will see, Heslop Harrison often didn't even provide examples of the plants he claimed to have found, sometimes offering the argument – even in the days before the conservation movement – that the plants were too rare to pick.

After reading how Raven knocked *Carex microglochin* on the head in his secret report in 1948, I looked the plant up in *Sedges of the British Isles* to see what Jermy, Chater, and David said about the plant in 1982. The text said baldly: 'The record for Harris is in error.'[31]

In the section of Raven's report dealing with the trip to Harris, there is a reference to Heslop Harrison's son-in-law, Willie Clark, that throws more light on the matter of 'the Clark paper that wasn't':

In this connection [Raven's conclusion that *Carex microglochin* never grew on Uisgnaval Mhor] there is one surprising fact that must be mentioned. This particular plant was apparently found, and certainly recorded, not by the Professor himself but by his son-in-law, Dr W.A. Clarke [*sic*], a botanist whose honesty has never been doubted and who seems indeed, to have been so disturbed by the suspicions surrounding the Professor's activities that for some years past he has dissociated himself from the Hebridean excursions. The only surmise that I can make to explain this fact is so unfounded that I do not think it is worth putting on paper.

What Raven believed, presumably, was that Heslop Harrison had somehow set up his son-in-law, either by planting *Carex microglochin* for him to find or even persuading him to pretend that he, Clark, had discovered it, so as to counteract the suspicion that was fuelled by all the rare plants having been discovered by Heslop Harrison himself.

After convincing himself that *Carex microglochin* had never grown on Uisgnaval, Raven applied similar arguments in his search for *Carex lachenalii*, another of Heslop Harrison's rare sedges. This time the clinching factor was that in the few other sites in Scotland where it was known, it was found above three thousand feet within a short distance of year-round snow patches. Here on Harris, the alleged site was below fourteen hundred feet, with little evidence of snow patches in the vicinity.

Raven rejected the third of Heslop Harrison's discoveries for different reasons. This was the sedge *Carex chordorrhiza*, which Heslop Harrison had described as growing in 'a patch of extremely boggy ground near Loch Stioclett.' Raven looked at very many patches of extremely boggy ground and failed to find it, but he accepted that this was not conclusive proof. Absence of evidence is not evidence of absence. But he did find a plant that appeared at first sight to be exactly the one Heslop Harrison had reported. 'It had the same far-creeping stem,' said Raven, 'with the same tufts of narrow leaves arising at irregular intervals from it.' It was only when he looked more closely that he realised that the plant he had come across was a submerged form of another sedge, *Carex limosa*, which he confirmed by finding other specimens of the same plant, with fruiting spikes that made it easier to identify. 'It is a mere surmise, though at least a plausible one,' said Raven, 'that it was this plant that was erroneously recorded from the same locality.'

It's interesting that Heslop Harrison himself was later to write on the difficulties of deciding between one sedge and another, and his views are referred to in the introduction to *Sedges of the British Isles*, where the authors say: '*Carex* is one of a number of groups of plants that present problems to the non-specialist. ... Some groups ... are distinct enough in minor characters [but] embarrass taxonomists by the vastness of their numbers and their slight differentiation from other parts of the populations (Heslop Harrison, 1953).'[32]

So within days of beginning his attempt to unravel the truth about Heslop Harrison's discoveries, Raven had found some persuasive evidence of misdeeds or at least incompetence. But he was not convicting his man yet. He next writes:

In spite of these conclusions, I returned from Harris with no evidence that I could reasonably expect anyone else to regard as conclusive. There still remained, of course, the Isle of Rhum, and it was there, if anywhere, that I had always felt that the solution of the problem was to be found. That feeling was largely justified by events.

On the last day of July 1948, after Arthur Clapham left, Maybud Campbell and Raven went to stay with a friend of Raven's, Walter Hamilton, who had rented a house near Arisaig, south of Mallaig, the port for boats to Rum.

One of Raven's oldest friends, Tom Creighton, had been invited by Raven to join him on the Rum part of his trip. They'd been at public school at the same time, and became firm friends at Cambridge. Creighton was rather a surprising choice – he was not a botanist and knew little about plants. He would certainly be much less help in corroborating anything suspicious that Raven might discover than half a dozen other friends or colleagues. But he had become one of Raven's closest friends, 'so close', says Faith Raven, 'that John's sisters got very fed up with it. He was always there, you know that feeling you have about people.' It's clear from an account Creighton wrote of their friendship after Raven's death that what Creighton supplied to Raven – and he to Creighton – on their trips together was an inexhaustible supply of laughter. He describes a journey they made by car through Wales in 1940 at the height of the German invasion scare, when they stopped to ask two ancient Welsh ladies the way. 'They looked us up and down,' says Creighton, 'brandished their umbrellas, broke into hilarious Cymrian cackles of laughter and said, "We aren't going to tell you. You might be the Germans. We're expecting them today," and marched firmly away. I swear the words quoted are *ipsissima* and that John laughed for at least [a] quarter of an hour.'[33] However motivated Raven was to get to the bottom of Heslop Harrison's activities, he cannot have faced with equanimity the prospect of crossing swords with a man almost twice his age and with a far greater reputation. It would have been comforting to have someone like Creighton along with him to share whatever vicissitudes he might face on 'the forbidden isle'.

A couple of years after Raven's death, Faith Raven had the idea of recording an interview with Tom Creighton about the events of the trip to Rum. Creighton himself died in 1993, but Faith Raven played me the tape in the low-ceilinged farm kitchen at Docwra's Manor, near Cambridge, the house she has lived in since 1954. While I listened, in my mind's eye I saw those two men, both in their early thirties, as they embarked with a mixture of excitement and trepidation on their risky quest. Describing the events of forty years earlier, Creighton gave an account that shows the fallibility of memory for unimportant facts and the absolute vividness of recall for those events that are emotionally important or intensely pleasurable.

Creighton had been in Berlin, and he sent Raven a telegram when he arrived in London that said: ARRIVING 8.40 AM TOMORROW, FULLY PREPARED. Creighton says: 'John showed the telegram to Walter [Hamilton] and said, "That means he's got a bottle of gin and a bottle of vermouth with him," which was perfectly true because in those days you could more easily get such things from the NAAFI in Berlin and bring them back by aeroplane than you could buy them in London.'

Creighton arrived on 2 August, and in the morning Raven tried to phone Lady Bullough to get permission for Creighton and Hamilton to accompany him onto the island. He encountered an unexpected snag. 'I found to my surprise,' Raven wrote in his report, 'that Kinloch Castle, Lady Bullough's monstrous residence on the island, had no telephone.' (Whatever happened to Rhum I?) Raven knew that if he and his two friends turned up uninvited in the boat from Mallaig, Lady Bullough's factor had instructions to turn them away, although Raven himself, of course, had permission to land, organised by Heslop Harrison. This was the very day Heslop Harrison himself was travelling to Rum, so Raven knew there wasn't much time to lose. He decided that the three of them would travel to Canna and take John Lorne Campbell's suggestion of entering Rum unobserved from the Canna side, with the help of Mr MacIsaac.

To get to Canna they would have to take the boat from Mallaig, which also called at Rum. And who should be on the boat but Heslop Harrison, expecting a visit from an unaccompanied Raven. 'I had to identify the Professor,' wrote Raven, 'whom I had never met but knew to be on board, and break to him the news that I should not after all be arriving till the 5th. I located him just in time and was instructed, somewhat peremptorily, to despatch a reply-paid telegram to Lady Bullough seeking her permission to bring Creighton with me on the 5th. Hamilton, from the distance, was not impressed by the Professor's appearance.'

For Creighton, this was all jolly good fun. The Inner Hebrides were 'a miraculous new world – the Highland seas, the birds, and the vast profusion of nature – having just flown out from blockaded Berlin.'

The three of them were deposited on Canna and found themselves a campsite on the east coast of the island, on a grassy flat of land beside the sea. 'It was very lonely, very beautiful, and, at that time of the year, very warm,' said Creighton. 'My memories are simply being happily together until the following morning, when a boat came by and a completely unknown Canna resident threw two lobsters, ready boiled, onto the turf beside us and went off. I've never understood the reason for this generosity, and the lobsters were extremely good to eat.'

On 3 August, Raven sent the telegram to Lady Bullough and the men spent the rest of the day on the island. Creighton then describes a strange visit to the house of John Lorne Campbell, the Laird of Canna and critic of Heslop Harrison's butterfly discoveries, who had suggested they approach Rum illicitly from the west:

He was a very distinguished collector of Highland folk song and Celtic lore, and he was also fairly eccentric. When we rang on the door of his house he came out and saw the three of us, Walter, John, and me, standing there. Without even mentioning our names or acknowledging our visit, he said, 'Oh, how nice that you could come. Now we can play a flute quartet.' The curious thing was that he did play the flute, and so did I. This made possible only a flute duet, if I had had my flute, which I didn't. He produced two flutes, but they were not of a system I had been taught to play, and John and Walter played no musical

Professor John Heslop Harrison in the 1930s.

John Raven, some years after his investigation of Heslop Harrison.

Right.
Alfred Wilmott, keeper of the British Herbarium and Heslop Harrison's rival in discovering new plants.

Below.
Maybud Campbell, a leading figure in the botanical community, described as 'Isabella' to Wilmott's 'Ferdinand'.

Heslop Harrison on a fieldwork trip, with one of his students.

A party of botanical field workers from Newcastle University on a trip to Rum. Heslop Harrison is on the far right.

Right.
Botanising on Rum, where Heslop Harrison 'discovered' many rare plants. He had been granted exclusive access to the island, and objected to other botanists encroaching on 'his' territory.

Below.
Heslop Harrison was accused of slipping specimens into his students' collecting bags while on field trips, to reinforce the acceptance of the plants as genuine.

Heslop Harrison's house in Birtley, Tyne and Wear. John Raven believed that Heslop Harrison had cultivated rare plants in his garden before taking them to Rum.

King's College, Cambridge, where John Raven was senior tutor.

Kinloch Castle on Rum, summer residence of the Bullough family, used by Lady Monica Bullough at the time Heslop Harrison made field trips to the island.

Kinloch Castle cost £250,000 to build – an enormous sum in 1900.

Left.
A bronze monkey-eating eagle in the Grand Hall of Kinloch Castle, brought back from the Far East by Sir George Bullough.

Below.
Plants on the Isle of Rum were believed by John Heslop Harrison to have survived the last Ice Age. The rare plants he claimed to discover there would reinforce that theory.

When John Raven visited Rum in search of evidence of Heslop Harrison's fraud, he visited this stream and found small groups of plants which he was convinced had been dug into the ground quite recently.

instruments at all, so no flute quartet took place. But the meeting was enjoyable, amusing, and like the Laird himself, highly eccentric.

The following day, Raven and his two friends were ferried across to the west coast of Rum, the opposite side of the island from Kinloch Castle, to Harris Bay. Here, Raven carried out his first test of Heslop Harrison's discoveries. 'I climbed Ruinsival,' he wrote, 'a hill of some 1,800 feet, in search of *Arenaria norvegica*, which the Professor had reported to be abundant there but which is elsewhere in the British Isles known only from Inchnadamph in Sutherland and Uist in Shetland. The record proved, as I expected it would, to be perfectly accurate: when a plant is reported as abundant there is good ground for supposing that it is there.'

While Raven was clambering up Ruinsival, Creighton, clearly uninterested in *Arenaria norvegica*, spent his time tramping in his heavily nailed climbing boots around the sandy margin of Loch Fiachanis, leaving, said Raven, 'an abundant supply of very evident footmarks.'

When Raven and Creighton returned to Canna in the evening, they found 'an obviously grudging affirmative from Lady Bullough'. The telegram said: NO ACCOMMODATION AVAILABLE. MUST BRING TENTS, FOOD AND ALL EQUIPMENT.

It was unfortunate that their journey to Rum was so disorganised; when they finally arrived on the island, they found they had left on the boat much of the very equipment that Heslop Harrison and Lady Bullough had pressed them repeatedly to bring. It was a worrying start to a potentially nerve-wracking visit.

7

'Quoth the Raven . . .'

I had only two sources for what happened on Rum during those two days in August: Tom Creighton's recorded memories, largely anecdotal and unbotanical, and John Raven's much more detailed account, at the heart of the report he wrote for Trinity, and both men are now dead.

In the last few days before he set foot on Rum, Raven must have been bolstered by the results of his visit to the Isle of Harris. However confident the inner core of British botanists was that Heslop Harrison was faking his results, there had been no guarantee that it would be easy to find him out. Now a case was beginning to emerge. But as Raven himself said after his trip to Harris, the evidence he had gathered would not be regarded by anyone as conclusive. Because of the timetable he had set himself, he now had forty-eight hours in which to come up with more convincing evidence to allow his colleagues to do what they were obviously itching to do: put the handcuffs on Heslop Harrison to stop him from betraying their beloved science of botany.

Tom Creighton, Walter Hamilton, and John Raven got up at 4:30 a.m. on 5 August to catch the boat from Canna to Rum, where Creighton and Raven would disembark; Hamilton would go on to Mallaig. The boat, run by MacBrayne on a circuit around the main Inner Hebridean islands, could not get right into the harbour at Rum, so a smaller boat, operated by Duncan

McNaughton, Lady Bullough's factor, had to take people off the steamer and carry them to Kinloch. This had the advantage of keeping away unwanted visitors, who could be refused transport on the smaller launch. Creighton described how they came to lose much of their luggage: 'Our landing was fairly dramatic. You could only be taken off the MacBrayne boat by the factor's boat, which went out to meet it. But we were legal immigrants, we had our papers. I only remember that we had large quantities of cooking pots and food on the boat, and we had to get off immediately under the ringed muzzle and horns of the rather large Highland bull who was tied up on the boat, being transported from one island to the other. Although it paid no attention to us, perhaps it was his presence which made us leave all our saucepans, cooking gear, and Primus stove, everything one needed to live, in fact, excepting a tent, underneath the bull's nose on MacBrayne's boat.'

It was only when the launch was fifty yards from MacBrayne's steamer and heading for Kinloch that Raven and Creighton realised to their horror what had happened. 'Those fifty yards were quite impossible to cross. Our boat was slower; we couldn't catch up MacBrayne. We waved, we shouted, we yelled, but MacBrayne wouldn't turn back, so we were condemned to spend our two nights on Rum without any cooking pots at all, and entirely and absolutely surrounded and devoured by the midge, because our antimidge precautions had been left on the MacBrayne boat too. I may say we never saw any of the things we had left again, of course.'

Once they had landed safely on the island and arrived in the small community scattered around Loch Scresort, Raven felt distinctly that the two of them were far from welcome. He put this down at first to the fact that, in spite of weeks of warnings, he had to confess that they had arrived without any food or cooking utensils. But then he observed the surprising fact that 'as soon as we had made it clear that we were not, after all, as our telegram had mistakenly suggested, friends of the Professor, the attitude of the inhabitants towards us underwent a sudden and total change. As our short stay on the island advanced, it became increasingly evident that the Professor was regarded by all the locals, including Lady Bullough herself, with considerable hostility and some suspicion.'

By 10 a.m. Raven was ready to begin his quest. He and Creighton reported to the professor's base camp and were told by Heslop Harrison to set out on foot and await the arrival of a car, bearing the professor and his party, at an appointed meeting place.

The car eventually arrived about 11.45 [Raven wrote], a late hour for a serious botanical expedition to begin its day's work. The party proved to consist of:

The professor

his wife

his niece

a certain Smith, whose botanical knowledge appeared to be confined to grasses

a girl of about 20, whose name was apparently Iolande and who was evidently there to gather instruction

Dr W.A. Sledge, of the Dept of Botany at Leeds University, a field botanist of note and evidently the honoured guest of the year.

This is a very Raven-esque passage. 'A late hour', 'a certain Smith', 'botanical knowledge . . . confined to grasses', 'evidently the honoured guest of the year' – each of these phrases is well chosen and some convey far more than the bald meaning of the individual words. Smith, of whom we never hear any more, clearly didn't say much at the time. Perhaps he made one or two remarks about grasses, then shut up, overawed by the much greater knowledge – and overbearing manner – of his leader. And the description of Sledge calls up images of Heslop Harrison assiduously drawing attention to particular discoveries in such a way that the rest of the group are fully aware of the mutual esteem that exists between the two 'great men'.

The group set off east from Kinloch toward a hill called Fionchra, which the professor claimed was botanically rich. Over the next few hours, Raven saw a series of plants that were certainly botanically interesting, although one or two seemed more like temporary residents than permanent inhabitants, under threat from the vicissitudes of everyday life on the island. One plant, an alpine *Thlaspi*, was too withered, said Raven, to be identified with absolute certainty. Another plant, a willow, that the professor attempted to show them was nowhere to be seen, and Heslop Harrison eventually suggested that it had been lost in a landslide. The group searched for another rare plant, *Lychnis alpina*, and were shown ten or twelve 'small, weak rosettes' of the plant at the foot of the cliff on a very steep little slope of loose, gravelly soil. Both Sledge and Raven were surprised that the plant was not to be found on the solid rock immediately above, but Heslop Harrison reported that it had been there previously, but was believed to have perished in yet another landslide. 'In its present locality,' Raven wrote, 'it was causing the Professor grave anxiety lest it should be trampled into destruction by passing deer – an eventuality which the extreme steepness of the slope seemed to me to make very improbable.'

Like so many observations in this story, the precariousness and occasional invisibility of some of these rarities were capable of two interpretations. If Heslop Harrison artificially created an atmosphere in which certain plants were difficult to find and constantly under threat, then the occasional absence of evidence need not necessarily have led to suspicion that they had not been there in the first place. On the other hand, rare plants must be rare for a reason. They *may* be constantly destroyed by landslides or trampled by cliff-climbing deer, in spite of the best efforts of the botanist to keep enough of them alive in their natural environment.

While the more botanically minded members of the party scaled Fionchra, Tom Creighton lay on the grassy slope beneath the cliff with Iolande, who was, perhaps, considered by the professor too frail to climb cliffs. At this point a misunderstanding occurred, due to Creighton's ignorance of the niceties of botanical practice. The incident he described had to do with Iolande's vasculum, a word with which Creighton was entirely unfamiliar but claimed to have thought referred to a part of the female anatomy: 'A young student stuck his head over the top of the cliff fifty feet above us and said, "Iolande, do you mind if I put something in your vasculum." She said, "No, I don't mind, but it's full of sandwiches." And because I didn't know what a vasculum was, I wondered how the sandwiches got where they were.'

After the first day's botanising, the members of the party returned to their accommodations – the Heslop Harrisons to their rooms in Kinloch Castle, while the rest of the party 'roughed it in a small house', in the words of Heslop Harrison's earlier letter to Raven. If staying in a small Rum house was roughing it, it is difficult to think of an appropriate description for Raven and Creighton's evening arrangements: a very small tent, with no food, cooking utensils, or any other creature comforts. But help was at hand, thanks to the unexpected attitude of the staff that Raven had noticed soon after landing on the island:

That evening Creighton and I called on Lady Bullough. While we were awaiting her arrival from her distant boudoir, her companion, Miss Rhodes, showed us a large raised map of the island and very kindly told us not only how best to climb Ruinsival but also where we might hope to find *Arenaria norvegica*. [The relief map of Rum is still displayed to visitors in Kinloch Castle.] I was glad – though I did not say so – to learn that my instinct had led me, the previous day, by the recognised route to the approved station. When Lady Bullough eventually appeared, not even the presence of a vast ivory eagle* brooding over Creighton's head could detract from the friendliness and charm with which she received us. We found it hard to believe that she resented the presence of strangers on her island so strongly as the Professor led us to suppose.

More good news followed. Creighton elaborated on the story of their visit to the castle on the first evening, describing a conversation that presumably took place after Lady Bullough had returned to her 'distant boudoir'. 'We were given a glass of whisky by the factor, and during the conversation we said we had no saucepans. The response was rather surprising. He said: "I can't let you take any with you because she might see you carrying saucepans, but if you wait, I'll bring them to you in the end." All I can remember for certain is that when we returned to our hungry camp with a certain amount of food but nothing to cook it in and waited for some time, a person – and it may have been Lady Bullough's French lady's maid – came very surreptitiously to the tent in which we were being devoured by the midge, with a very large brown paper parcel. Wrapped up in that parcel was a Primus stove, and two or three frying pans and a saucepan. The parcel was thrust under the flap of our tent, and I remember that our first action was to light the Primus stove in the tent to fumigate it of midges, and instantly the roof caught fire. So we then had no protection against possible rain, and no protection whatever against the midge. And so we survived for the next couple of nights. How we slept at all I cannot imagine, because the midge was worse than I have ever known. But I suppose we were

---

* The eagle is now in the Ivy Wu Gallery at the National Museum of Scotland in Edinburgh.

what is known in a cliché as young and resilient and put up with it.'

The following day, 6 August, any doubts Raven might have entertained about Heslop Harrison's activities were removed. Or it would be more correct to say that any *possibility* of doubt was removed, since Raven was clearly convinced of Heslop Harrison's guilt before the trip began. But the intention of the trip was to find evidence that would convince others in the profession – although nobody seems to have thought how such evidence might be disseminated and what might be done as a result.

On his last day on the island, Raven's investigations were made easier by the decision of Heslop Harrison's group to drive to Harris, the bay on the other side of the island where Raven and Creighton had landed illicitly two days before. The group were going to see the *Arenaria norvegica*, the plant that Raven had no difficulty in accepting as a genuine Heslop Harrison discovery. Raven couldn't really say that he'd seen it two days beforehand, so he just said he'd rather stay behind and see some of the other rarities on the mountain called Barkeval, halfway across the island. This might have been thought by Heslop Harrison as rather forward of him, but to Raven's surprise, the professor not only had no objection but produced a pencil and paper to give him directions for finding *Carex bicolor, Epilobium lactiflorum, Carex glacialis*, and *Erigeron uniflorus*. 'He told me, moreover,' Raven wrote, 'that I might collect a fruiting spike from a particular plant of *Carex bicolor*, and that, though I should find the *Epilobium* was past flowering, it was a perfectly simple plant to raise from seed.'

Raven obviously didn't know what to make of this burst of generosity on Heslop Harrison's part.

Nothing in the whole affair surprises me more than the readiness with which he allowed me to go without escort to examine his most precious discoveries. It might indeed be regarded as a powerful argument in support of his good faith. I am more inclined to think, however, that he simply dismissed me from serious consideration as an ignorant and incompetent fool – an attitude, I may say, which, even if it had been in my power to dispel it, I was at some pains to foster.

There is a third explanation. In addition to the possibility that Heslop Harrison's discoveries were genuine, or that they were fake but Heslop Harrison thought Raven too stupid to detect it, there is the possibility that Heslop Harrison was so convinced that he had covered his tracks in the matter that he felt no one, however clever, would be able to find any signs of fakery. After all, he knew that the plants were in the ground where he said they were, and how, after all, do you prove that something growing in a particular site was transported there rather than being a native plant? If that *was* the explanation for Heslop Harrison's insouciance, Raven was about to answer that question. 'Just as I was about to leave,' wrote Raven, 'Sledge, who had been standing by in silence, asked me whether I had seen *Juncus capitatus*.'

This was a rush whose appearance in Britain had so far been confined to the Channel Islands, West Cornwall, and Anglesey. Its alleged appearance on Rum was a remarkable extension of its previous sites – if verified.

I replied, truthfully, that I had only seen it in Cornwall. Thereupon the Professor, with a curt order to Iolande to come too, led Sledge and myself up the track beside the Kinloch River. After some 300 yards we approached a small rocky outcrop between the track and the river. Here, the Professor said, they had known for a number of years a very interesting and unexpected plant; and with evident pride he pointed out, on bare patches at the edge of the outcrop, a colony consisting of two unusually large and two small plants of *Polycarpon tetraphyllum*.

This was another little plant that had also only been found much further south until Heslop Harrison's team had discovered it on Rum. But there was something else unusual about this plant. Wilmott had been shown a specimen of *Polycarpon tetraphyllum* collected by Heslop Harrison's son-in-law, W.A. Clark, on an earlier expedition to Rum and had determined that it was a very unusual form of the plant, quite unlike the form found elsewhere in the British Isles.

This was important because if the plant had somehow extended its range from colonies that existed elsewhere in the British Isles – Cornwall, the Channel Islands, and Devon – a similar form of the species would have resulted. The fact that *Polycarpon tetraphyllum* on Rum was so different aroused Raven's suspicions that it had arrived on the island from a wholly separate source.

His suspicions were appropriately confirmed by the next event. Although the four had stopped only for a moment on their way to see the *Juncus*, Raven had time for a brief inspection of the *Polycarpon*. He noticed, and pointed out to the professor, 'the striking coincidence that out of the very middle of the largest plant of *Polycarpon* was growing a vigorous specimen of *Juncus capitatus*', one rarity intimately associated with another.

Now, the significance of such an event might be lost on a non-botanist, but it certainly wasn't lost on Heslop Harrison, who, according to Raven, issued 'the startled expletive, "Well, I'll be shot!"' He then hustled the small group on another two hundred yards or so to see *Juncus capitatus* in its approved station. As they walked up the path, Raven asked Heslop Harrison 'as guilelessly as possible' whether the *Polycarpon* was identical with the plant that he already knew in Devon and Cornwall. 'Well,' said the professor, 'according to Wilmott – if you can trust his determinations – it isn't. The plant has puzzled us a good deal, but we have eventually concluded that it must have been introduced with deer food.' Later Raven followed up this point and found very good reasons why this couldn't have been so.

Next, Raven was shown a flourishing colony of *Juncus capitatus*, and he asked if he could take a specimen. Heslop Harrison at first suggested that he take the *Juncus* growing out of the base of the *Polycarpon*, then changed his mind and gave Raven a plant from the colony, which he later pressed. But in a sudden change of plan, Heslop Harrison led the group back down the hill, where he uprooted the *Juncus* from the midst of the *Polycarpon*.

This, to Raven, did not seem the behaviour of a man who had discovered and was nurturing two specimens of rare plants:

I was considerably surprised that, instead of being naturally delighted that one of his most notable discoveries was extending its hold on the island, he seemed more concerned to exterminate it from its new foothold. And, since he then pointed out to Sledge and myself how well his specimen showed the rooting system of the plant, there is no possible doubt that that particular specimen will never reappear.

In Raven's mind, there was little doubt that the intimate juxtaposition of the two rare plants, and Heslop Harrison's reaction to it, showed that they had been imported to the island from some common location rather than having become established there naturally.

Armed with one piece of evidence, Raven and Creighton set off to find another, helped by the fact that the professor's party now set off for the other side of the island.

The most surprising of the professor's finds was the sedge *Carex bicolor*, entirely new to Britain. The site he indicated for this plant was close to that of another rare and un-British plant, *Epilobium lactiflorum*. With the help of Heslop Harrison's directions, the *Carex* was easy to find, about one and a half miles from Kinloch at the junction of two streams. Nine plants were scattered about a foot apart from each other on bare banks of gravel on either side of one stream. While Raven and Creighton looked at the location of this group, the same thought occurred to both of them. As Raven observed, 'If we had been looking for a suitable place in which to plant a remarkable botanical rarity, this was exactly the sort of locality – impossible to forget, and so bare that any plant on it was as conspicuous as its nature allowed – that we should ourselves have selected.' Then they looked at the plants one by one, and Raven noticed that one plant at least looked as if it had been recently 'dibbled in' – dug into the gravel with a trowel. It also appeared half dead, as if it hadn't taken root as it should.

As Raven continued his methodical examination of the plants, he must have scarcely been able to believe his luck. A close inspection of the other plants revealed growing evidence of the likelihood that they, too, had been cultivated elsewhere, away from Rum, and planted in this location.

The first fact that struck me [wrote Raven], was that one tuft of the *Carex* contained, sprouting from its very midst, as *Juncus capitatus* had sprouted from *Polycarpon*, a vigorous plant of *Poa annua*. Now *Poa annua*, though it does sometimes occur on mountains (as, for example, around the summit cairn of Ben Lawers or near the Scottish Mountaineering Club's hut on Ben Nevis) is not a plant that I should normally expect to find in a Scottish corrie [a circular hollow in a mountainside]. It is primarily, though not exclusively, a weed of gardens, paths and roadsides. Moreover I noticed, on looking still closer, that it was accompanied in the middle of this particular tuft of sedge by another plant of precisely the same predilections, namely *Sagina apetala*. I drew Creighton's attention to this extraordinary fact and asked him to witness my examination of the remaining 8 plants. As might have been expected, the two very young plants were free from contamination; but of the mature plants two others besides that already examined proved to contain *Poa annua*, while yet two more contained *Sagina apetala*.

Like a prosecuting counsel eager to make sure that the jury doesn't miss the significance of every bit of evidence, Raven then wrote:

> Of the 7 mature plants, in other words, 5 were accompanied by weeds associated primarily with gardens. The only two that were uncontaminated were, first, the plant that I have already mentioned as being more than half dead, and, second, the largest and healthiest tuft . . . which may well, in my opinion (since it was with this plant that the Professor had, in my preliminary conversation with him, shown himself most delighted), have received more careful grooming than the rest.

Obviously, whatever his botanical experience told him about the improbability of *Poa annua* and *Sagina apetala* being found naturally in the area, Raven needed to be sure that the two species entwined in the tufts of the *Carex* weren't also in evidence in the surrounding gravel: he searched the gravel banks around the area, 'examining, I believe, every single plant that had found a foothold on them' and failed to find any examples of either.

There was one final suspicious discovery before Raven and Creighton left the site. About seven feet from the rare sedges, Raven found, to his astonishment, 'a small plant of the by now only too familiar *Juncus capitatus*'. He was convinced that this plant would not occur naturally in this particular site and wrote, 'The only question concerning its presence there is whether it was deliberately or accidentally introduced.'

Raven removed a fruiting spike of the *Juncus* and took two specimens of *Sagina* from a tuft of *Carex*. In the roots of one were a moss, a liverwort, and a pellet of soil. These specimens would help, Raven hoped, to provide some material confirmation of his suspicions.

In Raven's current frame of mind, every detail of the scene could be used to feed his doubts about Heslop Harrison's data. The location of the *Carex* that he had just seen was not the site from which it was first reported. There were no longer any examples of the rare sedge at the original site, which was on steep terraces on the other side of the corrie. According to Heslop Harrison, the plants must have been exterminated by landslides or deer. The site Raven and Creighton visited at the junction of the two streams also presented some risk to the plants:

> It is interesting to note that in its present station too – and this is perhaps another reason that led to the selection of that station – the entire colony could be very easily washed away in a spate. There is no doubt at all that the gravelly banks on which it grows are periodically submerged beneath the swollen waters of the burn. There is thus, if the new colony, like the old, fails to maintain its hold, a convenient excuse always ready to hand to account for its disappearance.

Although the report was written a few days after the events, it reproduces the growing certainty in Raven's mind of Heslop Harrison's guilt. The use of 'selection' in the passage above says it all: the site was *selected* by Heslop Harrison to be the station for the *Carex*.

This growing certainty led Raven's standards as prosecuting counsel to slip a little below what would be allowed in a court of law when he presented an interesting sidelight on the story of the discovery of *Carex bicolor* based entirely on hearsay. But it's a juicy bit of hearsay that, if true, gives another firm tap to one of the nails in the coffin of the professor's reputation.

I have recently heard, at second hand, a very strange story relating to the first discovery of *Carex bicolor* in its original station. This story was told me by Professor A.R. Clapham, who had himself just heard it from Mr R.B. Cooke. Cooke used, in the early days before his suspicions were aroused, to accompany Professor Heslop Harrison's party on its Hebridean excursions, and he had in fact witnessed the first finding of *Carex bicolor*. The Professor had apparently set out on that particular day's walk armed with a trowel – an instrument that he had seldom carried before but for which he accounted on this occasion by the statement that he intended to 'dig beetles'. When Cooke, with Dr W.A. Clarke, was some 200 feet higher up the slopes of Barkeval than the Professor himself, there was suddenly a loud shout from below and the Professor came running up the hill with a plant in his hand, saying, 'I've got a new sedge'. Cooke had to content himself with the sight of the specimen that the Professor had already dug up. The Professor was not, as might have been expected, at all eager to show the rest of the party the colony from which his own specimen had come.

Raven then pointed out that the Professor's account of this episode differed somewhat from Cooke's account as reported to Clapham and passed on to Raven. In his *Journal of Botany* paper, Heslop Harrison described his expedition as following the course of a tributary that ran between a series of ledges or terraces: 'These terraces are sparsely clad with various sedges, grasses, etc., and on one of them I was fortunate enough to detect a colony of an unknown sedge from which I collected five specimens. On our return journey, after much trouble, in spite of the care taken over bearings, in refinding the plants, I pointed them out to other members of the party.'[34]

So far, then, Raven and Creighton had hunted down three of Heslop Harrison's rarities. There were still several more to be sought before the day was out, and they set off to look for *Epilobium lactiflorum*, a plant said by Heslop Harrison to grow around a hundred yards away from the *Carex*. There it was, a dozen plants or so, growing from a mossy spring and looking natural and healthy. But Raven was suspicious that a similar spring a few yards away, which provided an even more suitable site, showed no trace of the plant.

There was one further incriminating aspect of the two sightings of *Carex* and *Epilobium*. Neither was accompanied by what Raven called congeners, plants you would expect to see in the neighbourhood. Every plant has its congeners, which flourish in the same type of soil and climate. These two plants were Arctic alpines, and with *Epilobium lactiflorum*, for example, Raven would have expected to find at least one other similar plant, *Epilobium alpinum*. But there was no trace of it.

Raven and Creighton set off in search of a last few plants – with rather inadequate directions. Ironically, it turned out later that they had been within a few yards of *Carex capitata*, a plant shown to Sledge by Heslop Harrison a couple of days before. When first discovered several years earlier, according to Heslop Harrison, there had been six or seven plants but now there was only one. He was concerned that the other five or six had disappeared because of the avidity of some unknown and unauthorised botanist, but to Raven this was 'a conjecture which, since he himself had assured me that without minute directions I should never find the plant, did not impress me as very plausible'. The other reason offered by the professor for the plants' disappearance, as paraphrased by Raven, was 'a lack of botanical discrimination on the part of some sheep that had recently been introduced to the island'.

Raven was unconvinced by the story of the missing plants:

I myself would surmise – though this is nothing but a guess – that there had never been more than the one plant; and it even

seems to me not impossible that the plant was the very one
Professor Butler had been shown in 1947 in the plot of ground
by the Kinloch River, having been transplanted into the present
station during the Professor's preparatory visit to the island in
June of this year.

As Raven and Creighton began to look for two more plants on
their list, *Carex glacialis* and *Erigeron uniflorus*, Raven realised that
the task was hopeless:

I had been told simply that they grew together on the shoulders
of Barkeval just below a large boulder that 'looked like a squashed
tank'. The whole shoulder proved to be covered with innumer-
able boulders, any one of which, to a lively imagination, might
have resembled a squashed tank. As I had been warned that there
were only 3 plants of the *Carex* and hardly more of the *Erigeron*
– facts, incidentally, of the utmost significance in themselves,
since no genuinely British alpine is quite as scarce as that – I
determined not to waste more than two hours on an obviously
hopeless search and instead dropped down again to Kinloch to
pay another visit to the *Polycarpon*.

Raven obviously wanted some time alone with the plants that he
had seen in rather a hurry and under the nose of Heslop Harrison
that morning. He hoped that a more leisurely inspection would
throw up further clues as to their origin. The *Polycarpon* site was
actually within sight of the Heslop Harrison base camp, but Raven
knew that the professor's party would probably still be on the other
side of the island visiting Harris. He was also throwing caution to
the winds, as he realised how much evidence he had gathered
already of Heslop Harrison's misdeeds. 'I came to the conclusion,'
he wrote, 'that I should at this stage run the risk of my peculiar
activities being observed by the Professor.'

At this point Raven made a discovery that he described as 'perhaps the most significant of them all'. Growing out of one of the *Polycarpon* plants was a diminutive flowering plant that he had never seen before, although, he added – baldly but accurately – 'I am acquainted in the field with every known species of Scottish plant.' He thought it might be another of Heslop Harrison's discoveries, called *Cicendia pusilla*, not because he recognised it but because it was one of the few British plants he'd never seen.

This plant was to prove the most significant discovery because of its value in reinforcing the hypothesis of which Raven was growing more and more certain.

Raven and his colleagues were convinced that Heslop Harrison was cultivating a number of rare plants somewhere away from Rum, probably in his own garden at Birtley or on some other plot under his control. If he had been doing this successfully over a number of years, it is likely that, assuming the garden included a range of other plants of botanical interest, or even just weeds, some of these other plants would have mingled with the rarities as seedlings or seeds in a way that would not have been evident at the time they were dug up to be transported to Rum. The *Juncus* that sprouted from the *Polycarpon*, and the *Poa* and the *Sagina* that emerged from the *Carex*, could have been carried inadvertently from Birtley when Heslop Harrison set off to transport the two rare plants. But while Raven's suspicions about the origins of the intruders were plausible, they were not conclusive. *Poa* and *Sagina* might have existed already on Rum, although he had failed to find any examples. And *Juncus* did already exist on Rum, but at some distance from the *Polycarpon* (and in any case the *Juncus* was already under strong suspicion of having been imported).

But the diminutive flowering plant that Raven had never seen before was to provide the most persuasive evidence of all that the

Polycarpon, at least, could not have been a natural inhabitant of Rum.

> At this point [said Raven], I shall have to anticipate. Soon after my return to Cambridge I spent a large part of a morning, with Dr Catcheside and Dr Richards, in an attempt to identify the plant . . . The only conclusion we reached was that it was not in fact a member of the British flora at all. And this conclusion was confirmed very soon afterwards by Mr H. Gilbert Carter, who recognised it at once. It proved to be *Wahlenbergia nutabunda*, an abundant weed of the Canaries, which is sometimes cultivated in Botanic Gardens in Britain, where it tends to survive by seeding itself, but elsewhere in Britain is entirely unknown . . . The conclusion became quite irresistible. The *Polycarpon* had been, like *Carex bicolor*, deliberately planted; and whoever it was who had planted it had been so careless as to leave in the soil surrounding its roots the seeds of two other plants that were to betray its origin.

In the space of a day, by tramping up hill and down dale and using his finely tuned observational skills, Raven had turned the uncorroborated suspicions of the pillars of the botanical community into a near certainty. It was definitely time for a drink, and Raven and Creighton accepted an invitation from Lady Bullough's factor, Mr McNaughton, to take a dram of whisky with him and his wife, and with Bryce the butler, down at Kinloch Castle.

In the course of 'a most friendly and interesting conversation', Raven made two more minor discoveries. Heslop Harrison had told Raven earlier in the day that seeds of the *Polycarpon*, an unusual form of the plant, might have been introduced with deer food. The conversation in the servants' quarters revealed that deer food was indeed imported to the island, from the River Plate in

South America, an even more unlikely site for *Polycarpon* than Rum, and certainly a place where *Polycarpon* had never been found.

The second discovery emerged from the fact that McNaughton and Bryce knew Heslop Harrison quite well, from his visits to the island over more than ten years. 'In the opinion of Bryce,' Raven wrote, 'which was pretty clearly shared by the more cautious McNaughton, "the professor kept something up his sleeve – either a butterfly or a plant – to discover every year." '

As the two young men made their way back to their tent, they must have been very pleased with the day's work and were probably dying to talk over the large number of detailed observations that Raven, in particular, had made. But that conversation was to be delayed a little longer as they met two of Heslop Harrison's party, who invited them back to the base camp for a cup of tea. Here, wrote Raven, another

friendly and interesting conversation ensued, in the course of which, while I was engaged in botanical technicalities with the Professor and Sledge, Smith and Iolande reported to Creighton their startling discovery, on the shores of Loch Fiachanais under Ruinsival, of the recent marks of a large pair of heavily nailed climbing boots. Creighton, having accurately described the course of his walks on the past two days, neither of which, of course, took him anywhere near Loch Fiachanais, closed the conversation by remarking, again perfectly truthfully: 'No doubt it was somebody who landed without permission on the far side of the island. I understand it's a common enough practice.' I do not think that even then they had the slightest suspicion of our actual motives. At all events . . . the farewells of the entire party, and particularly of the Professor himself, were as cordial as could have been desired.

It is unlikely that Raven and Creighton could have kept a straight face for very long after leaving the professor's base camp. If those at the base camp were startled by distant howls of laughter as the two friends finally made their way back to their tent, they would have had absolutely no clue as to what had caused one of those shared moments of hilarity that so cemented the friendship.

8

'Is Such a Thing Done?'

In the recording Faith Raven made of her conversation with Tom Creighton, he summed up his own view of what had happened on the Isle of Rum, based on his personal, if uninformed, observations and also, obviously, on the report written by his close friend: 'All this evidence, for what it's worth, has made it absolutely convincing to a total amateur, an absolute layman like myself. I have not the slightest doubt that things are as [Raven] suggested, that Heslop Harrison had been planting plants, that the whole thing was an utterly pathetic, very elaborate confidence trick by a person who should have been an extremely distinguished scientist and failed to be so, and was trying to redeem his loss of name in a way that was much more – to me – pathetic than criminal. But, then, I wasn't a committed botanist. If you were a committed botanist, you had to get the record straight, you had at any cost in human terms to expunge error from the botanical record, and that this error existed and was properly expunged I have no doubt whatever. I am the only person with a completely neutral attitude who saw all the evidence, and I was able to judge it.'

Of course, in one sense, Creighton was far from neutral. If you have become close friends with someone whose abilities you admire and whose judgment you trust, and if, further, you have no informed basis for judging whether his conclusions are correct or not, you are hardly going to doubt those conclusions. What

Creighton's statement reflects, I think, is what Raven believed in private, the strong version of what he wrote in his more measured report.

Raven summarised his views in the report to Trinity College by grouping his conclusions under two headings: psychological and botanical. This piece of writing makes a fine peroration, and the psychological conclusions are the stuff of closing speeches to juries by prosecuting counsel:

> There seem to me to be three possible explanations that might account for the facts – namely: –
>
> 1. That the Professor is himself the victim of a practical joke. This explanation, though it has the merit of being the most charitable, has the outweighing defect that it not only fails to account for, but actually conflicts with, a number of the facts recorded.
>
> 2. That as Dr Jekyll he plants the specimens which as Mr Hyde he later discovers.
>   This, though I find it an attractive explanation, is certainly somewhat romantic – which is no doubt why it appeals to me. And since there is, I think, no doubt that Mr Hyde remembers – rather than independently discovering – what Dr Jekyll has done, it may well be also psychologically absurd.
>
> 3. That the Professor is deliberately indulging in the most culpable dishonesty in order to secure for himself an immediate reputation and an immortal place in the annals of British botany.
>   This is, I fear, the only explanation that really fits the facts. I am only too well aware that I, who have no scientific training whatever, am bringing a very grave charge indeed against one who, besides being a Professor of Botany is

also a Doctor of Science and a Fellow of the Royal Society. I need hardly say that unless I were sure of the ground upon which my charge is based I should do no such thing. The Professor is clearly a man of very humble origin: he is in fact, I have been told credibly enough, the son of a miner. He is obviously able and almost equally obviously ambitious. He is indeed – as Sledge, Creighton and I all independently felt – the type of which dictators are made. Most unfortunately he is, I believe, so constituted that he can never rest content with the honest fruits of his genuine ability. There are, in the history of Scottish field botany, several names that might well fire the imagination of any modern British botanist . . . The form that the Professor's ambition has eventually taken is, I believe, the determination to add his own name to this notable list. Rhum, of course, provided the ideal scene for the fulfilment of this aspiration. Exceptionally mountainous, geologically diversified and rich, it had the two supreme advantages that it was hitherto almost unknown and that ingress to it was notoriously difficult. By the simple expedient of exaggerating this already notorious difficulty the Professor has contrived, for some ten consecutive seasons, to keep the field entirely to himself and his few invited guests. I am firmly persuaded that the combination of his own peculiar psychology with the opportunities offered him in the Hebrides in general and in particular in Rhum has led him into the astonishing aberrations that I have described. Whether on that account he should be exposed and disgraced I find it difficult to decide. I cannot help hoping not, since, for all his eccentricity, he treated me throughout with a rather touching, if usually gruff and peremptory generosity. But one thing seems indisputable – that in the interests not only of truth but also of the reputation of

> British science it is essential somehow to discover what
> plants and what insects he has either completely fabricated
> or else deliberately introduced into the Hebrides.

Raven's botanical conclusions were simple and clear. *Carex bicolor* should be struck off the British list, into which it had been admitted in 1941; *Polycarpon tetraphyllum* should be struck off the Hebridean and Scottish lists; and ten further plants should also be struck off the various lists onto which they had found their way in previous years, thanks to Heslop Harrison's announcements, 'unless or until they are ever found again in different circumstances in the Hebrides'. Finally, as a result of his trip to Rum, Raven was convinced that at least two of the Professor's attributions were accurate, *Arenaria norvegica* and *Thlaspi calaminare*, and he had no hesitation in saying so.

From Rum, Raven travelled the few miles south to Arisaig to collect his thoughts and plan the next stages in the process of pinning down Heslop Harrison's guilt. Then, on 10 August, he began the long journey south to Cambridge on the train that travelled from Arisaig to Glasgow along the meandering line that took in some of the most beautiful scenery in the Highlands. On board, he wrote the first draft of the detailed report that described his activities and observations on Rum while they were still fresh in his mind.

Back in Cambridge he had two important tasks that were essential follow-ups to his investigations. He had a number of specimens with him that could be shown to more senior botanists to make sure that his own conclusions were correct. He also had to make his findings known to Heslop Harrison and give him a chance to provide an alternative explanation for the discoveries that, to Raven, clearly pointed to Heslop Harrison's deliberate misdeeds.

One of the people whose expertise would help confirm some of the specific points Raven had made in his report was, of course,

Wilmott. Whether or not he played a part in initiating Raven's trip is unknown, but there is a letter in Wilmott's file in the Natural History Museum archives that suggests Raven kept him closely informed. It was written on Wednesday 11 August, the day after Raven returned to the Lodge at Christ's College, his family home when he wasn't at Trinity.

> Dear Mr Wilmott [Raven wrote respectfully, from the junior to the senior man],
>
> I got back yesterday from my wanderings in the Hebrides, where I had an exceptionally interesting time. I succeeded in carrying out your commissions on Rhum, and have secured for you, besides a very fine sheet of *Arenaria*, a fairly large selection of other interesting plants. They are not, I fear, very well dried, because a damp tent 6′ by 4′ × 4′ is not the ideal place in which to cope; but I am pretty confident that you will find them interesting.

The phrase 'your commissions on Rhum' could mean two things. It could mean simply that Wilmott had wanted Raven to bring back plants to augment his collection of Hebridean flora; the one mentioned, *Arenaria norvegica*, Raven had concluded, was one of the professor's genuine discoveries. Or it could be that discussions with Wilmott and others beforehand had led to the specific task of incriminating Heslop Harrison by acquiring samples of the suspect plants and bringing them back to the British Museum for expert assessment. The correspondence that had ended so acrimoniously in 1945 may well have lit the fuse. Whatever the significance of the phrase, Raven went on in his letter to seek at least an hour of Wilmott's time during the following week to discuss his findings.

At the same time that he was organising the Natural History Museum visit, Raven contacted a number of other scientists for help in confirming some of his findings. In Cambridge, he visited the School of Botany and handed over to two botanists, David Catcheside and Paul Richards, the moss and the liverwort he had found in the middle of one of the tufts of *Carex bicolor*, hoping to find that they were alien to Rum and therefore further evidence of 'planting'.

Then there was the other matter that needed to be attended to. Whatever happened to his report – and at the time it was unlikely that he thought it would be buried the way it eventually was – Raven knew that he had to give Heslop Harrison some indication of his findings and some opportunity to reply, even though he felt there was no other explanation than one of his psychological conclusions.

There was, however, a small complication. Raven felt more than a tinge of guilt, since he recognised that he had gone to Rum under false pretences, fully determined to find evidence under Heslop Harrison's nose that would discredit him. He would have to find a way to deal with that issue when it arose, as it was sure to once Heslop Harrison realised what had happened.

Faith Raven told me that one of John's friends pointed out to her how much Raven enjoyed the chase. She had not appreciated that side of him, but once she thought about it, she saw their own courtship in the same light: 'If a girl was someone else's girlfriend, John would want her, and I suppose that's what happened with me.' Many years later a friend, Nick Jardine, describing a series of lectures Raven gave on ancient Greek botany, said:

Throughout the lectures John succeeded in communicating a specific enthusiasm, an enthusiasm which gave life to all his interests, classical, philosophical and botanical, the enthusiasm of the hunt. Sometimes the quarry was a person. In the opening lecture the hapless Sir William Thistleton-Dyer, F.R.S. . . . is hunted down. With unholy glee the superficiality of his scholarship and the absurdity of his diagnoses are gently but remorselessly exposed.[35]

The correspondence with Heslop Harrison that followed the Rum trip has the characteristics of the final stages of a chase, rather like, to change the metaphor slightly, the way a fisherman plays a fish on the end of his line once it has taken the fly. In his first letter, Raven started with a gentle tug here and there, and then, as Heslop Harrison responded to each tug, Raven pulled a little bit harder in each of his replies.[36]

Here's how Raven's first letter to Heslop Harrison begins, apparently more in puzzlement than in the certainty that he felt, written on Friday 13 August, 'having consulted Prof. P.W. Duff on the laws of libel', Raven noted in his report.

Dear Professor,

I am writing to thank you very warmly for the generosity with which you treated me during my short stay on the Isle of Rhum.

I deeply appreciated the opportunity of seeing so many interesting plants, and I am extremely grateful to you not only for personally showing me the large majority of them but also for giving me the directions that enabled me to see the two others that I succeeded in finding on Barkeval.

I may have mentioned to you that my College was so generous as to give me a grant of £50 towards my expenses in the Hebrides. This places me, of course, under the obligation of making a report to the Council of the College, an obligation which I cannot evade and which is causing me considerable embarrassment. I am very sorry to say that I was gravely perturbed by a certain amount that I saw in Rhum . . .

In the light of what we know about the antecedents of the trip, this paragraph is, of course, very economical with the truth.

Raven then set out the key observations he made: the *Poa* and *Sagina* growing in the roots of the *Carex bicolor*, the *Juncus* in the midst of the *Polycarpon*, and so on. At this stage, he let Heslop Harrison believe that he, Raven, accepted the possibility that 'some unauthorised and irresponsible botanist had been at work on the island'.

The day after he sent off this letter – surely with some trepidation – Raven received a letter from Sledge, to whom he had put his preliminary suspicions on the boat back from Rum to Mallaig.

Sledge's letter suggests that he had no idea, until Raven raised the matter, that there was anything untoward about the professor's discoveries:

Your information was not only unexpected but particularly unpleasant as it came so soon after my cordial parting from H.H. As I told you, I had found him a very agreeable companion, anxious to show me anything I wanted to see, placing no restrictions on my collecting and free from any suggestion or secretiveness about any of his discoveries. I noticed too that when personalities with whom I knew he was more or less strongly antagonistic cropped up in conversation he did not display the violent disagreement or scornfulness which I might well have expected. I had therefore come to like him and it was a great shock to me to hear of your observations on the *Carex bicolor* plants. I feel now that with the most charitable feelings it is still impossible to escape the conclusion that there has been deliberate planting of the colony. And so all sorts of doubts are raised about the others. If I were in your position therefore (but I am glad that I am not!) I should most certainly not withhold any of the incriminatory evidence from the report.

I find two things surprising about this letter. One is that Sledge, as a reasonably experienced and senior botanist, had heard nothing of the rumours that had apparently been circulating for some time. Recall that in Raven's report he said that 'a number of the best qualified of British botanists came to view this lengthening list of Hebridean rarities with growing suspicion'. Either Sledge was not in informal contact with 'a number of the best qualified of British botanists' or Raven was really writing about suspicions that had been discussed among Wilmott and his circle and not much more widely than that. The second surprise is that, for all the protestations of friendship between Sledge and Heslop Harrison, Sledge was ready at the drop of a hat to believe that his 'friend' had committed a serious fraud on botanical science. And this was on the basis of a conversation on the MacBrayne steamer and without the benefit of Raven's detailed report. There are enough mysteries in this story for me not to bother too much about these two small ones, but the most likely explanations to me are (a) that the rumours of Heslop Harrison's misdeeds were less widespread than Raven's report implied, and that they hadn't reached Sledge, a man at a different university and working on different plants; and (b) that this letter was another indication of the way – still in evidence fifty years later among the few people still alive who knew him – that Heslop Harrison could charm some people at the same time he antagonised others.

The flurry of activity continued back in Cambridge. Raven revisited the School of Botany on Saturday, the 14th, to hear the verdict of Catcheside and Richards on the liverwort and moss samples. To Raven's disappointment, they told him that the plants were so widely distributed in the British Isles that their occurrence on Rum would not be unusual enough to ascribe an alien origin to the *Carex*. A trip to the Cambridge Botanic Garden the following day, a Sunday, enabled Raven to see for himself a sample of *Wahlenbergia nutabunda*, the plant from the Canaries that he had discovered growing out of the middle of a large *Polycarpon* plant on Rum. To check the theory that Heslop Harrison had taken this from the Newcastle Botanical Gardens, where it could have established itself as a weed, Raven then enlisted his father in the chase.

Charles Raven, by now Vice Chancellor of Cambridge University, wrote a letter on 15 August to an acquaintance, Dr D.H. Valentine, Reader in Botany at Newcastle. It left little room for doubt about the accusations being made against Heslop Harrison: 'I have seen the evidence,' Canon Raven wrote, 'and it is plainly in the interests of scientific truth that it should be probed.' He asked Valentine whether *Wahlenbergia, Polycarpon*, and *Juncus capitatus* grew in the Newcastle Botanical Gardens, and if so, could he send some samples? 'Of course,' Canon Raven wrote, 'we realise that *Polycarpon* and *Carex bicolor* may probably have been grown privately, but the presence of intruders with them on Rhum strongly suggests that they were grown elsewhere and planted in.'

On that same busy Sunday, John Raven visited a mineralogist, Professor Cecil Tilley, and gave him the pellet of soil he had removed from under the *Carex* in the hope that analysis might show some ingredients that were not a natural part of Barkeval soil.

The following day, Raven left for London. In the appendix to his report, describing the events and correspondence that followed his trip, he described his visit to Wilmott on Monday 16 August: 'On my way through London I took the opportunity of visiting the British Museum (Natural History) and there, as the obvious repository for such material, I left all the specimens the collection of which is recorded above.'

I may be reading too much into too little, but it seems to me that the phrase 'as the obvious repository for such material' might be part of an attempt to play down the fact that Wilmott had been instrumental in planning the Rum trip. The fact is, according to Raven's letter to Wilmott, that Wilmott had asked him to do something on Rum, Raven had done that something, and he had now arranged, a week in advance, a meeting with Wilmott to report on his success. In any case, Raven wrote:

Mr A.J. Wilmott confirmed all those of my determinations which might conceivably have been questioned. With him I also compared my specimen of *Polycarpon* with every single specimen in the Museum's world herbarium. It was not precisely matched by any, its combination of much-branched inflorescence and narrow acuminate sepals being apparently most unusual. But it resembled most closely the specimens of *Polycarpon tetraphyllum* from Malta and Greece.

In addition to providing plant samples to Wilmott, on his return from Rum, Raven had sent specimens of *Arenaria norvegica* and *Thlaspi calaminare* to Professor Tom Tutin. On 18 August Tutin wrote to acknowledge their safe arrival and added, 'I collected *Carex bicolor* in Switzerland and am interested to see that it appears to be very easy to cultivate. My plant has made a lot of growth in the past month.'

On 23 August, Canon Raven received a reply from Valentine, another senior botanist surprised by the allegations against Heslop Harrison.

The facts that have been discovered are astonishing; the whole situation is regrettable, and I agree that it should be looked into, but I am most reluctant to take any steps myself. I have met Harrison a good many times and have botanised with him in this neighbourhood, so that I know him quite well, though I have never talked with him in any detail about the Hebrides. I'm sure you will understand that, in the circumstances, I prefer to abstain from any action.

He went on to say that the plants mentioned were certainly not in the experimental greenhouse at Newcastle, and very probably not in the Botanical Garden. 'Harrison has a large garden at his private house at Birtley, and I think he grows his plants and does his experimental work there. I have never visited the garden and know practically nothing about it.'

For a solid fortnight since he had returned from Rum, Raven had indulged in – and stimulated – a whirlwind of activity: letters and samples in all directions, trips around Cambridge laboratories and botanical gardens, no let-up even on Sunday. During the same fortnight he had written his first letter to Heslop Harrison and now awaited the reply. It was two weeks before it arrived, but the delay was a result of Heslop Harrison's absence, not of any period of reflection or contemplation of the issues. Deceived, perhaps, by Raven's calm reasonableness, Heslop Harrison replied in the same vein, with only one or two mild criticisms of Raven that can hardly have surprised the recipient:

30/8/48

Gavarnie,

The Avenue,

Birtley,

Co. Durham

Dear Raven,

I am sorry that there has been a delay in my reply to your letter which I received this morning. I had to return home last week owing to the illness of my mother but she unfortunately died before I reached home. Under the circumstances I could not go to Newcastle until today.

It is a very great pity that you did not inform me of your observations before you left the island for then I could have discussed the facts with you in person and given you the necessary information for which you ask.

Here are my comments:

1. I too have seen *Poa annua* and a *Sagina* (which I considered offhand to be *S. procumbens*) with *Carex bicolor*. This caused me no surprise for, on Barkeval, and Mullach Mor on the opposite side of the glen, *Poa annua* abounds at high levels

amongst the colonies of Black-backed Gulls and Manx Shearwaters. These colonies you have failed to note. Moreover, you do not seem to be aware of the facts that, not infrequently, plants of *Poa annua* and other species are to be found on pony and deer droppings at all levels.

2. *Sagina apetala* is very rare on Rhum and was encountered only on Barkeval and Kinloch Glen, our earliest record being from the same area on Barkeval in 1937.

3. I have never seen *Juncus capitatus* away from the Kinloch Burn, but its occurrence with the *Polycarpon* is not surprising as it was originally noted on Rhum on the other side of the stream exactly opposite the *Polycarpon* rock, i.e. about a dozen yards away. I have never seen the plant on Barkeval, but I see no reason why a casual seed should not have blown up the corrie.

4. I attach no importance to *Polycarpon* which, as I told you, I consider to have originated with food imported for wintering deer. It has occurred sporadically along the road since we first saw it in 1938. Actually, there are odd plants just in front of the house in which the party lived but I did not consider it worth the trouble to point them out to you. However, I showed them to Sledge.

5. Of *Cicendia pusilla* on Rhum I know nothing: none of us has ever seen it there.[37]

6. The original colonies of *Carex bicolor* were higher up the corrie on the broken ground to the left as you ascend. Let me know if I can do anything further to assist.

Yours sincerely,

J.W. Heslop Harrison

Surprisingly, Raven was a little shaken by this reply. 'I felt at first reading,' he wrote, 'that some – though not all – of his arguments carried a certain weight. But on more mature consideration I felt justified by September 5th in posting [an] Open Declaration of War.'

The Open Declaration of War is not quite as Open and Warlike as one might have expected, knowing what Raven knew and with the answers he had to all the professor's points – or had gathered from friends and contacts in the few days after receiving Heslop Harrison's letter. In particular, it stopped short of a direct accusation of fraud aimed at Heslop Harrison. Instead, it left open the door, again, for 'some unauthorised botanist or party of botanists' to have committed the deeds that Raven spelled out in more detail.

In his letter, Raven wrote: 'I have given a great deal of thought to the questions involved and my opinion is by now as definite as it ever could be.' As Raven saw it, there were two issues in Heslop Harrison's letter that needed to be dealt with: the idea that the other plants mixed up with the rarities could be dismissed because they grew elsewhere on the island; and the status of *Polycarpon*. Heslop Harrison seemed to be saying now – in Raven's view for the first time – that *Polycarpon* was not native to the island but had been introduced with deer food. Since writing his first letter, Raven had discovered that the '*Cicendia*' growing out of the *Polycarpon* was actually *Wahlenbergia nutabunda*. There could be no doubt that the presence of *Wahlenbergia* among the *Polycarpon* established that the *Polycarpon*, previously reported by Heslop Harrison as native to the island, had been introduced there from some site of cultivation rather than from deer food. In what was clearly a provocation to Heslop Harrison, Raven suggested that a report ought to be written for a leading botany journal recording the presence of the *Wahlenbergia* and mentioning it in the context of the introduction of *Polycarpon* to the island. What was particularly provocative was that Raven suggested that he should write it himself. As we have seen, one of Heslop Harrison's trigger points for suspicion was the thought that someone else was trying to take credit for something he had discovered.

As far as *Carex bicolor* was concerned, with *Poa* and *Sagina* in its roots, and *Juncus capitatus* not far away, Raven emphasised that it wasn't so much the presence of these plants on the island he took as suspicious – enough people had told him, if he didn't know it already, that *Poa* could be all over the place, establishing itself very easily – as that in the immediate vicinity of the *Carex*, these were the only specimens of these plants. 'This remarkable coincidence,' Raven wrote, in his reply to Heslop Harrison, 'seems to me to be wholly unaffected either by the occurrence of quantities of *Poa annua* around the bird colonies on the neighbouring hills or by the occasional appearance of such annual species on pony or deer droppings.'

After referring to the 'unauthorised botanists' hypothesis, Raven closed his letter by mentioning that he had soon to submit his report to the Council of Trinity and that he would be grateful if Heslop Harrison could send him any suggestions that might help explain the facts listed. In his report, Raven adds one further comment on Heslop Harrison's letter, when he says that he cannot see any purpose in the addition of Heslop Harrison's paragraph six, about the supposed location of the original colonies of *Carex bicolor*, 'unless it is by way of preparation of a last line of defence'.

Declarations of War are often followed by Hostilities, and Raven's was no exception. His letter, posted on the fifth, reached Heslop Harrison on the sixth (how reliable the post was in those days!) and stimulated an immediate reply. This demonstrated some defensive tactics that Heslop Harrison might have thought were rather clever, although to a disinterested observer they look a little like panic. Heslop Harrison stated in his letter that 'there is already in the printer's hands a short article dealing with the points you raise. In that article I list all the plants I have seen with the *Carex bicolor* colony you were shown.' He quoted a passage that he said was from his field notebook for 1943 to prove that he, too, had seen a number of plants with the *Carex*, including *Poa annua* and *Sagina*. The only explanation for writing a paper now, five years later, about those observations was to pre-empt anything Raven might do along similar lines, as Heslop Harrison admits later in his letter.

In quoting from his notebook, Heslop Harrison then mentioned that the *Carex bicolor* plants were infected with galls, small growths created by some insects in which to lay their eggs. The particular insect in this case was a gnat, *Pseudohormomyia granifex*, and Heslop Harrison enclosed a report from an entomological journal – written by him – of this gnat's first appearance in Scotland, on the *Carex* on Rum.

Having prepared a defence, Heslop Harrison then turned to the attack: 'You refer to the supposed acts of a botanist or group of botanists. I know that one such group, to put it mildly, does not look upon our Hebridean work with favour.' (He is presumably referring here to Wilmott and his co-workers, clearly not in Raven's mind as possible planters of the rare plants on Rum.) 'However, my opinion of human nature must be higher than yours. I cannot conceive that any human being would be vile enough to sabotage our work in the way you suggest.'

Supporters of Heslop Harrison might well point to this sentiment as an indication that he was as appalled as any good scientist by the possibility of fraud or tampering with data. But the response might also suggest the canny realisation that to seize on the excuse proffered him would mean accepting that something wrong had been done, whereas he persisted in denying that anything was amiss with the data.

Heslop Harrison continued:

In my letter I wrote that it was a pity that you did not communicate with me on Rhum. I now feel that my statement was too mild. It was your duty to do so, more especially when, contrary to the assurance you gave Lady Bullough, you took away a rare plant considered by you to be *Cicendia pusilla*. Further, I should remind you of your strict undertaking that you would not communicate the locality of the rarer Rhum plants to anyone. This is coupled with the fact that you are not entitled to publish anything about plants shown to you under such a pledge. In any case pure courtesy demands that no one should deal with plants forming the subject of another's researches. That is why I wrote the paper to which I referred above.

In a spirited peroration, Heslop Harrison then displays the first suspicion that there is more to Raven's inquiries than a simple botanical quest for information: 'In my opinion your exaltation of trifles, capable of simple explanation by any ordinary person, indicates that consciously or unconsciously you are influenced by something quite unknown to me.'

As Raven pointed out in his comment on this letter in his report, if the observations he had listed were 'trifles, capable of simple explanation by any ordinary person', why had Heslop Harrison bothered to dash off and send to the printer an article evidently designed to explain those observations? Raven then added, 'In any case I have, by inducing him to write such an article, achieved the first of my objectives. What my right course will be after its publication must depend, obviously, on its contents. But for the moment at least the letter which I posted on September 9th may, I hope, have closed the correspondence.'

John Raven was a complex character. Several people mentioned that he was physically sick with nervousness before giving lectures. And yet here he was severely criticising – and being criticised by – a senior figure in the botanical establishment, with a coolness and sureness of touch that raised Heslop Harrison to higher and higher levels of incandescence. But his confidence was no doubt bolstered by doing this with the connivance and support of other senior botanical figures, such as Wilmott and his associates.

Now, he hoped, having done the right thing and given Heslop Harrison the opportunity to present some sort of explanation, one that he, Raven, was capable of dismissing, he could move on to the next stage: deciding, with the help of friends and colleagues, what to do about the whole business.

Raven's letter of September 9 gave oleaginous thanks for Heslop Harrison's help, expressed his relief that Heslop Harrison's article would deal with the problem points, and told of Raven's hope that the article would dispose of his misgivings once and for all. But as a letter designed to 'close the correspondence', it did nothing of the sort.

I place no value on graphology, the science of reading character, personality, and even the future from someone's handwriting, but Heslop Harrison's final letter in this exchange (though not his final letter ever to Raven – that comes later in the story) could only have been written by a man at the height of his fury. The handwriting races across six pages, with occasional crossings-out and missing words, in a nonstop flow of aggrieved invective, aggravated, perhaps, by a state of ill health to which he refers at the beginning of his letter: 'For the past three weeks I have been suffering from severe heart trouble and am quite unfit for anything. I therefore write this final letter to state plainly what I think of your various actions.'

'The past three weeks' would be the period since he had read Raven's first letter, which, presumably, came right out of the blue. If this was the first direct accusation of fraud – even if only implicit – that Heslop Harrison had ever received, and if he really had spent the last decade in undetected fraudulent activities, such a letter could well have caused a violent shock to his system, making him 'quite unfit for anything'. More charitably, however, he had also suffered the death of his mother at about the same time, and such an event may also have plunged him into ill health.

At this stage in the correspondence, Heslop Harrison appeared to have abandoned much of an attempt to deal with the issues and just poured out his anger. He clearly resented what he saw as a series of breaches of good manners and botanical practice committed by Raven, a resentment he seems to have nursed since a failure on Raven's part three years beforehand to acknowledge receipt of some plants for which he had asked (to which I can find no reference in Raven's papers). The full letter (see Appendix, page 243) contains complaints about the manner of Raven's initial approach to him, his failure to specify correctly who was in his party, and his arrival on the island on a different date from the one he had originally indicated. In all, there are eleven itemised points of complaint against Raven, a veritable litany. Two items conclude, 'Is such a thing done?' Heslop Harrison also hints about a certain unnamed person on the island who had a personal objection to Raven and who seems to have embarrassed the professor and his wife by blaming them for his presence.

The letter closes with a statement that, however much else was disputable in this correspondence, Raven would have had to admit was almost the truth: 'Finally, as a result of my realisation of all these facts, I have come to the conclusion that you came to Rhum prepared to find fault, and therefore, quite naturally, but without the slightest justification, managed to do so.'

It was clear to Raven that there was no point in continuing a correspondence that had led to ever-lengthening and ever more irate replies from Heslop Harrison, so he sent a brief note by return of post to close it: 'Your letter reached me this morning. To reply to it in detail would clearly serve no useful purpose. I write, therefore, simply to acknowledge its safe arrival.'

By the same post as Heslop Harrison's long letter came a letter from Professor Vero Wynne-Edwards, editor of a journal called the *Scottish Naturalist*, enclosing the article that Heslop Harrison had said was 'in the hands of the printer'. Wynne-Edwards told Raven that he felt it 'necessary to take the most careful advice before accepting any of this author's contributions for publication', a remark that confirms that suspicion of·Heslop Harrison already existed among a wider group than Wilmott and his colleagues. Because of Wynne-Edwards's suspicions, he had consulted Sledge, who in turn had suggested he send the article to Raven because, in Sledge's words, 'I think the article has been written for a specific purpose and is not as innocuous as it might appear.'

In his report to Trinity, Raven included a paragraph copied verbatim from Heslop Harrison's article, having presumably returned the complete text to Wynne-Edwards. The article was titled 'Observations on the Flora of the Isle of Rhum' and the extract runs:

> Para 3. On May 29th, 1943, a survey was made of the plants growing near or with *Carex bicolor*. These included . . . At intervals since there have been noted growing with, or amongst, the sedge odd specimens of *Sagina* spp, *Stellaria media* and *Poa annua*, no doubt originated from seeds washed amongst the plants when the burn was in spate. On the roots of the *Carex bicolor* plants numbers of galls of the rare British gall-gnat *Pseudohormomyia granifex* were found; this species was new to Scotland. Of the rarer plants recognised as introduced, *Cotoneaster simonsii, Polycarpon tetraphyllum* . . . etc. were again encountered . . .

Following a reference to this extract, Raven then explained for the benefit of the Council of Trinity the significance of the article:

> The whole is, as Sledge had immediately recognised, a somewhat unskilful attempt to deprive my observations in advance of their potential sting. I replied at once to Professor Wynne-Edwards with a brief account of the circumstances that had given rise to the article, and added that, though I could not therefore honestly advise him to publish it, I none the less hoped

that he would see fit to do so, because I might then feel justified in publishing a reply.

The day's correspondence, in fact, very nicely presents the whole crux of the matter. Professor Heslop Harrison is perfectly justified in his denunciation of my breach of professional etiquette. Though I am pretty certain that without the Professor's help no conclusive evidence would ever have been forthcoming to determine the status of the plants in question, it would still be with a very uneasy conscience that I actually published observations that were only made possible by accepting that help. At the same time the letter from Professor Wynne-Edwards seems to show how important it is that the problem should be publicly aired and settled. By suppressing my observations I become an accomplice to what I firmly believe to be deliberate fraud. If the Council, either as a body or as individuals, would consent to advise me in this dilemma, I should be sincerely grateful; and if, either as a body or as individuals, they feel that I should overcome my scruples and publish a bald statement of the botanical evidence that I have recorded in this report, then I should be genuinely and profoundly relieved to have their moral support.

Three other pieces of evidence fleshed out the botanical case in the following ten days. First, Raven had sought advice about the gnat *Pseudohormomyia granifex* that the professor had reported as affecting *Carex bicolor* with its galls and had claimed as an insect new to Scotland. There had been only two recent reports of this gnat in England, one from Cheshire and the other, Raven was astonished and pleased to find, by Heslop Harrison from Birtley. This seemed too good to be true. A piece of evidence supplied by the professor himself showed that the *Carex bicolor*, which Raven believed had been cultivated in the professor's garden in Birtley, was infected by a gnat that had been reported previously by Heslop Harrison himself from his garden in Birtley.

To get an expert view on this, Raven had sent a soil sample to a Cambridge entomologist, Mark Pryor, and he quotes Pryor's verdict, warning that too much should not be read into the coincidence: 'If the creatures exist in Cheshire,' he told Raven, 'they probably exist all over the country, if anyone takes the trouble to look for them. I suppose that if there is a secret Caricetum* at Birtley and it is infected with *Cecidomyids* [gall-gnats], one would expect the Rhum plantations to be infected too.'

(The actual letter from Pryor begins 'My dear Holmes', a playful reference to Raven's sleuthing.) Raven went on to point out that, if the garden also contained *Wahlenbergia nutabunda* as a weed, 'there could hardly be any further argument', though he added, 'but apart from the practical difficulty of gaining admittance, I am not disposed to press my investigations quite so far'.

The second piece of information that emerged in late September related to *Polycarpon* and the question of whether Heslop Harrison had originally reported it as an introduction to Rum, as he insisted in his letters to Raven. Now, on referring to Heslop Harrison's *Flora of Rhum*, published in May 1939, Raven found the following:

Polycarpon tetraphyllum L.

A single plant found growing in a rock crevice along the Kinloch Burn, Rhum; a new county record.

Raven showed this reference to two experienced botanists, both of whom confirmed his view that it could only be interpreted as referring to a native species, rather than to an introduction.

The third significant piece of data with which Raven finished his report was the result of a visit he made to the Cambridge

---

* Area where sedges were cultivated.

Department of Mineralogy on 2 October. A Mr McLoughlin had analysed the pellet of soil taken from the roots of the *Carex* on Rum and found two types of minerals: olivine, which reflected the underlying rock where the *Carex* was found, and quartz, which didn't. The obvious explanation would be that the quartz had washed down from higher up the hill, but as the entire hill was of olivine, this didn't seem likely. McLoughlin concluded that the quartz had been artificially introduced, although Raven admitted that because quartz is a ubiquitous mineral its existence in the *Carex* soil could, by itself, carry little weight. Still, added to the rest of the information, it strengthened the case against *Carex* being a native of Rum.

In the eight weeks since his return from Rum, Raven, with the help of a band of botanists, entomologists, and mineralogists, had built up a formidable case against Heslop Harrison. But as Raven's plea to the Council of Trinity showed, he was not at all sure what to do next.

9

The Aftermath

If John Raven had been a writer of fiction, he could not have created a more circumstantially compelling case against Heslop Harrison than the twenty-one closely written pages of his report to the Council of Trinity. Raven's report displays an ability to use his skills in the way a writer of fiction should, to involve the reader, keep him guessing, maybe even sensationalise a little.

To the Council of Trinity College this might all have seemed very persuasive. Few if any were botanists; several were scientists but from very different disciplines; the rest – the majority – were historians or philosophers or members of other humanities faculties.

Fifty years later, it was impossible to find out what really transpired when the council was presented with Raven's report. The college archivist could find nothing relevant to the matter, and in talking to people who might know something, I came across the merest hints, surmises based on second- or third-hand information, about what might have happened. Tim Clutton-Brock, for example, the Rum deer expert, told me: 'My memory is that John said that the people in Trinity suggested that Heslop Harrison should be put on formal academic trial in front of his peers at the Royal Society, and then they debated that and decided that it shouldn't go that way.'

It is actually difficult to see what the council could have done in any official sense. They had no control over Heslop Harrison or any official connection with the botany establishment or any other body that might have been able to reprimand or castigate him, or publicise his misdeeds to a wider audience.

By late 1948, the circle of people who knew about Raven's report was wider than the Council of Trinity. Enough people had been consulted by Raven about parts of the story for the word to get around that something was afoot. And since those people already had reason to doubt some of Heslop Harrison's earlier discoveries, there was a sense that the matter could at last come to a suitable and destructive climax. Some of this was expressed in a letter to Raven from Vero Wynne-Edwards. He had turned down Heslop Harrison's hurriedly written article and accompanied the rejection with the intimation that the pages of the *Scottish Naturalist* were not open to him. He thanked Raven for his comments on the article and continued:

> I have so lately been drawn into this deplorable business, and am so innocent of the facts, compared with a number of other people, that I do not feel I am the right person to take the responsibility for making a public exposure. I have an idea that A.J. Wilmott may already have organised some plan for doing this, and believe you should get in touch with him if you have not already done so.

It is difficult to know from this quite what Wynne-Edwards had heard. Since Raven and Wilmott had been in close contact over the preceding weeks, it may be that the idea of some kind of publication was already being hatched between them. Wynne-Edwards went on:

> You must feel most reluctant to make the first move also, but if you are satisfied that your observations on the deliberate introduction of *Carex bicolor* put the matter beyond reasonable doubt,

and would withstand cross-examination, then you would do a public service by stating in print that such a deliberate introduction had been made. Heslop Harrison's name need not be mentioned. I would gladly publish such a note in the *Scottish Naturalist*.

This letter was typed on 22 September on Wynne-Edwards's university department letterhead. But he obviously had second thoughts about the 'public service' he was recommending to Raven, because within a day he had written to him again, a handwritten note on *Scottish Naturalist* letterhead, perhaps from home. He had clearly been worrying about the effect on Heslop Harrison of any exposure of the matter.

If we are to look for clues as to why this story had never been fully explored in the fifty years since it happened, the most important one lies in this letter. Not because Wynne-Edwards himself had much to contribute: he was very much a bit-part player. But his second, hurried letter to Raven encapsulates a particularly English concern: the harm that would be done to the perpetrator if his misdeeds were publicised.

I do not see how HH's name can be kept out of any exposure of the fact that *Carex bicolor* was planted in Rhum with intent to defraud the world. And if he is accused of it, it will be a great personal disaster to him, which could end in his resignation from the university, if not worse. No one wants such things to happen and in fact there is no doubt that a public exposure should be Since writing to you last evening, I have thought more about the subject. I do not see how avoided if possible.

Wynne-Edwards actually wrote 'should be avoided at all costs,' then crossed out 'at all costs' and replaced it with 'if possible'. He went on: 'I think he should be approached personally and privately, and this course persisted in until there is no longer any doubt that it is a waste of time, or until it has some effect. He will have great difficulty publishing any of his papers, I believe, which is a good thing.'

Raven could certainly have told him that the idea of approaching Heslop Harrison 'personally and privately' would have been unlikely to have any effect, at least on the basis of his correspondence so far.

At this distance in time, and without the passion for natural history that motivated the participants in these events, I find Wynne-Edwards's letter a little melodramatic. For me the interest in the story is not the earth-shattering nature of the planting of *Carex bicolor* on Rum, it is that, as we will see, the Rum events were part of a much wider pattern of activity. We have already seen doubts about some of Heslop Harrison's other botanical discoveries, and we will hear more about his work in other areas. But for Wynne-Edwards, the *Carex bicolor* incident alone was clearly a heinous crime. And yet, heinous as it was, he decided that it should not be allowed to blot the career of a man who, although near retirement, still had years of active work ahead of him that might lead to further such deeds.

For the next twenty years, such admirable but short-sighted concern for the feelings of others prevented anyone from doing very much about Heslop Harrison's research activities, and until today the professional scientists who knew of the story have kept pretty quiet.

In fact, Raven didn't act on the advice in Wynne-Edwards's second letter. He decided to write a paper for a scientific journal. This would be the true test of whether he had written fiction or not, a public account of his observations that, he knew, would be read closely by Heslop Harrison and his friends, who would seize on anything that could be disputed. At least, that is what he must have thought, although the actual sequence of events was rather different.

I find it surprising that the journal Raven chose for his paper was *Nature*. Contrary to what its name suggests, *Nature* was then, as today, the leading journal for the publication of major papers across the whole field of science, including physics, astronomy, chemistry, geology, and biology, as well as botany. Raven's contribution wasn't a major scientific paper, admittedly: he wrote it as a letter to *Nature*, a form of communication that was used to convey preliminary announcements of important or interesting data. Nevertheless, the form in which the paper was cast, describing the alien nature of a couple of plants on a distant Hebridean island, was still an unlikely use of *Nature's* columns.

I suspect that there must have been a certain amount of behind-the-scenes assistance to get the piece into *Nature* at all. I can't imagine that such a letter, arriving in the post from a classics don at Cambridge, would strike *Nature's* subeditors as of immediate and pressing interest to its readers. But someone like Wilmott, for example, a leading figure in British botany with a senior post at the British Museum, might well have been able to put a word in with the journal's editor.

The final piece was written on 17 December and, with surprising speed, appeared four weeks later, on 15 January 1949. It was titled 'Alien Plant Introductions on the Isle of Rhum'. Unlike a scientific paper, a letter to *Nature* could relax the rules a bit, and this meant that Raven could write in the first person. This he did, and in a piece of Raven-esque effrontery, he wove Heslop Harrison's name into the very first paragraph in an apparently innocuous context.

A number of very rare and interesting plants have been reported in recent years from the Isle of Rhum, in the Inner Hebrides. Though Lady Bullough, the owner of the island, generously gave me permission to camp there for longer, I was unfortunately unable to spend more than three days on Rhum, and my knowledge of the flora is therefore very fragmentary. But thanks to the kindness of Prof. J.W. Heslop Harrison, who was staying on the island at the same time, I was enabled to see some at least of its most interesting plants. Of these I shall be especially concerned in this note with two only, namely, *Polycarpon tetraphyllum* (L.) and *Carex bicolor* All.

The rest of the letter presents a summary of Raven's findings on Rum, augmented by the fruits of his research after he returned and his foreknowledge of Heslop Harrison's answers to some of his points. For example, knowing from his correspondence with Heslop Harrison that the professor would argue that he had always said that *Polycarpon* was an introduction, Raven quoted Heslop Harrison's original report of its discovery, which says nothing of the sort.

While Raven made no direct accusations against Heslop Harrison, the two references to the first reports of the plants on Rum were to Heslop Harrison papers. To anyone who cared – and that was likely to include only a select subset of *Nature*'s readers – the letter was saying: 'This man discovered these two rare plants on Rum, and I now say that they were "planted".' As we might say nowadays, 'Go figure.'

In a sense, the *Nature* letter marked a natural end point to Raven's part in the story. Although for years afterward the events formed a topic of conversation among Raven's botanical friends, he had little or nothing to do with Heslop Harrison or his dubious reports subsequently.

But meanwhile, what of the professor? His ability to respond rapidly when threatened – the Raven correspondence by overnight post, the pre-emptive article 'in the printer's hands' within hours – should have been expected to lead immediately after Raven's *Nature* letter to some kind of red-hot missive in the journal editor's in box.

For a long time nothing happened. *Nature* received no reply or comment from Heslop Harrison or any of his allies in the months following Raven's letter. Then, in February 1951, Raven, now a Fellow of King's College, Cambridge, received this letter:

The Editors of NATURE present their compliments to Mr J.E. Raven, and beg to enclose a communication entitled ALIEN AND RELICT PLANTS IN THE HEBRIDES, by Dr. J.W. Heslop Harrison. They will be glad to know whether Mr Raven wishes to make any comments upon this communication for publication or otherwise. The Editors have struck out the greater part of the first paragraph, which would not be appropriate for publication in NATURE in any event.

Accompanying this letter in John Raven's papers is a typescript of a piece by Heslop Harrison. I assume that this is what the editors of *Nature* sent to Raven, but if so, it is difficult to see quite what they were worried about in the first paragraph, which merely says: 'My attention has been drawn recently to a letter by J.E. Raven which appeared in *Nature, 163*, 104 (1949), and dealt with alleged alien plants on the Isle of Rhum.' But what is interesting about this paragraph is what it says about the lapse of time since Raven's *Nature* letter. What had been going on here?

It is, of course, a well-used device to show critics how beneath you they are by saying that your 'attention has been drawn' to their criticisms. Can Heslop Harrison really not have known about the *Nature* letter for two years? I imagine it would have been the talk of the Senior Common Room at Newcastle within hours of publication. But who would tell the professor? As we will see later, he said that his wife knew about the letter before he did, but she made sure that he was not told. On several later occasions, Heslop Harrison stuck to his story that he found out about the *Nature* letter only in 1951, and there is no reason to doubt him. Later he said that a student had drawn his attention to Raven's letter, and this is all too believable. While his colleagues and acquaintances might have been too cowed by his powerful personality and the knowledge of the letter's significance, some hapless student may well have come across it and mentioned it to his professor.

The piece Heslop Harrison wrote for *Nature* was a refinement of his argument that some at least of the rarities on Rum had escaped from the disused greenhouses and conservatories of Kinloch Castle. This excuse disposed of the *Polycarpon* and the *Wahlenbergia*, even though he had originally reported *Polycarpon* as if it were a native. As far as *Carex bicolor* was concerned, he stuck to his story, building up a case based on his belief that the terrain in Rum *is* similar to other localities where *Carex* is found, and on the statement that there were many more specimens of *Carex* on the island than the few plants that Raven reported, and that *Poa annua*, which Raven cited as an importation along with the *Carex*, could be found there in abundance.

The second paragraph of this letter might have been the one that triggered the objections of the *Nature* editors. It contains various remarks about Raven, such as 'It is difficult to imagine . . .' and 'He fails to mention . . .', which may have fallen outside the permitted language in the view of editors who wrote to people in those days 'begging to enclose' communications. In any case, whatever the editors wanted to do to Heslop Harrison's letter before publication, he was not having any of it and withdrew it, preferring to publish a more expansive and outspoken reply in his newsletter called *Occasional Notes*.

This same newsletter contains another example of Heslop Harrison intemperateness. When I spoke to David Allen, the botanical historian, he told me of an incident in which a young amateur botanist, Donald Young, felt the sharpness of Heslop Harrison's tongue. Young had the task of abstracting botanical articles from the scientific press for *Watsonia*,[38] the journal of the

Botanical Society of the British Isles. He wrote what he thought was a reasonable summary of Raven's letter.[39] Then, in *Occasional Notes No. 6*, just after his attack on Raven's letter, Heslop Harrison stuck the knife into Young's abstract: 'In *Watsonia*, ... there appeared what purported to be an abstract of the letter discussed above. Part of this "abstract" is purely bogus, and the remainder even more tendencious [*sic*] than the original letter . . .'

Young was apparently so upset by Heslop Harrison's fury at what he had written that he stopped abstracting botanical articles for good.

There is one more piece of correspondence in Raven's file that helps to round off the Rum story. It's a letter from Ramsbottom's successor at the British Museum, George Taylor, enclosing an extract from a letter written by a colleague and occasional co-author of Heslop Harrison's, an amateur botanist called Randall Cooke, the 'R.B. Cooke' mentioned in Raven's report as witnessing the first finding of *Carex bicolor*. Taylor's letter is the only reference I have come across that attests to the possibility that the Newcastle people were organising themselves to carry out some kind of concerted action against Heslop Harrison. Taylor wrote, 'Cooke, I know, will want to be in at the kill but he wishes to prepare his brief in collaboration with Clark.' At this stage, presumably, with so much becoming public knowledge at least in the botanical community, Clark had decided to assist in the task of dealing with his father-in-law, although there is no evidence that 'the kill' ever took place. The extract from Cooke's letter gives, in his own words, the story that Raven quoted at third-hand in his report:

I am interested in what you say about John Raven. When H.H. made his 'find' in May, 1941, he gave me three plants of *Carex bicolor* with very well-developed fruits. One I pressed and the other two I tried to grow in the garden, but they, unlike other Rhum plants except Orchids, refused to thrive. I think the pressed one shows fruits too well developed for the time of year and the altitude at which it was said to be found . . . I did not see them 'growing' in 1941, but was told that there were a number of plants over a small area. On my last visit to Rhum in 1946 Clark showed me the place where he had seen them in 1941, but we could find no trace of even a single plant. Although I did not see them being planted I feel as sure they were planted as if I had seen it. Now I expect the *Carex* has been tried in a number of other places on Rhum . . . Between 1941 and 1946 I was only shown one plant in the ground. Probably in 1941 it was thought not wise to let me see them, being at that time, more of a gardener than Clark. In 1941 it was on the second day of our visit (that was repeated on several other occasions). He took a trowel with him, to dig for spiders he said (both very unusual). I walked with him till near the place where Clark saw them and only left him when the digging started. At this point Clark and Miss H. would be about 200 ft. above us and I had almost joined them when we heard a shout and saw H.H. hurrying up towards us, and as he got nearer he called out, 'I have found a new *Carex*'. Of course, at this point our climbing was slow as we were examining all the rock outcrops.

Willie Clark is a shadowy figure in this story, but as Heslop Harrison's son-in-law and colleague he must have been very close to the truth of what went on. That he was happy to join with Cooke in preparing a brief against Heslop Harrison is a strong indication that, for all Heslop Harrison's excuses to Raven, even his closest colleagues were convinced that *Carex bicolor* at least was a 'plant'. When Professor John Morton, Heslop Harrison's friend, gave me an account of his view of the whole affair, he added a detail about Clark's involvement in the story that I had heard from no one else, and it was consistent with the idea that Clark had become shocked and disillusioned with Heslop Harrison's ways.

'The rumours as I recall them,' Morton told me, 'were that on one of the Prof's expeditions to the Hebrides, when his son-in-law, Dr Clark, and another well-known naturalist from the northeast, Dr Carter, accompanied him, the Prof showed him several specimens – picked flowers – of new discoveries, which he claimed to have just made. Clark and Carter became suspicious because the cut ends of the specimens were going brown, as if they had been kept in water for some time. They apparently confronted the Prof with this and there was a great row. I can just imagine the Prof's reaction to such a suggestion. Both Clark and Carter returned home and did not participate in further expeditions to the Western Isles. Who spread the rumours I don't know, but rumours of this sort spread like wildfire, for the scientific community is a very close and inter-connected community.'

As we have seen, those rumours persisted for forty years or more, long after Raven closed his files on the matter and went back to his classical studies. And because they were rumours rather than open allegations, Heslop Harrison was able to pursue his interests doggedly into old age with few ill effects other than, perhaps, a gnawing resentment of Raven's accusations that was little different from the grudges he harboured against dozens of other people with whom he had crossed swords in the past.

Those rumours kept alive two main questions, discussed among botanists whenever the allegations came up, which have still not been answered. The first, of course, is 'Was Raven right?' and the second, which I'll turn to now, is 'If so, why did Heslop Harrison do it?'

IO

Seeds of Doubt

If Heslop Harrison was the only distinguished scientist ever to have been accused of faking results, it would be an uphill task to establish the case against him 'beyond reasonable doubt'. But it is a sad fact that from time to time discoveries are reported in the scientific press that turn out later to have been faked. And the exposure of these fakes involves stories that have much in common with the scientific life of John Heslop Harrison. Someone believes in a theory, expects to be able to extend its significance with more research, fails to get the results he needs, and decides to manufacture them.

Tim Clutton-Brock helped me understand the steps that could have led Heslop Harrison to fake results, and, most important, to believe at some level that it was justifiable to do so. 'From my perspective as a scientist,' Clutton-Brock said, 'for anyone to do what is suggested for Heslop Harrison, he would have to cross a major bridge. If you take something like the Piltdown Man, whoever did it was clearly a crook and knew he was a crook and knew just what he was doing. Now, if you're going around "planting" plants, you must fall into that category. It's very difficult for a scientist to do that – they are trained that the data are terribly important. I suppose the sort of thing that Heslop Harrison might have done, which would fit more in my mind, would be that he could have said, "Well, it would be evidence that *Carex bicolor could* occur here if it persisted here." I could vaguely see how you might say, "Okay, well, I'll plant it here and see if it does survive." And then you could conceivably say, "Well, good heavens, look what I've found – I didn't plant that one. Maybe that's a wild one," and that turns into "That must be a wild one." For it to be credible to me, I'd either want to believe Heslop Harrison *was* a crook, which is what John Raven believed, or I'd want to create some sort of downward slide into believing that [the plants were native] without sharp interruption, and it's quite difficult to create that, to cover the chasm of actually planting and introducing.'

That difficulty of belief in the villainy of a fellow man – a scientist, 'one of us' – that Clutton-Brock, with all his English sense of fair play, experienced may be one reason the Heslop Harrison affair has never been publicly acknowledged: people have never really wanted to believe that Heslop Harrison had stepped into the chasm.

In a book about another type of deviant science, Felix Franks quotes a leading British observer of how science and scientists work, John Ziman, on the subject of fraud in science:

> Ziman suggests that scientists can, and often do, maintain two different sets of standards. They 'will intrigue for political ends like any Jesuit, and can be as lordly as any hospital consultant in the control of their juniors. They can deceive their wives, fiddle their tax returns, drive drunkenly, live beyond their means, feed parking meters, beat their children, and otherwise live as antisocially as anyone else when the occasion demands.' Ziman goes on to say: 'I have the very strong impression that these traits are repressed within research itself.'[40]

I have to say that I think Ziman has got things 180 degrees wrong. It is *because* they are capable of deceiving their wives, fiddling their tax returns, and beating their children, for all sorts of social rewards and reinforcements, that a little bit of data manipulation or even the regular and persistent manufacturing of evidence presents them with no problems. They're only human, after all.

I have chosen a couple of cases of alleged fraud by scientists – there are many to choose from – because they highlight several factors which have operated in the Heslop Harrison affair: the conviction that a scientific theory is correct leads you to believe that it's all right to fake a few results; the situation in which large numbers of the members of a community of scientists know about dubious work and do nothing about it; the very human tendency to avoid criticising the eminent or the powerful because you believe in their ability to use their power against you; the role of ordinary human drives for fame, money, or sex in affecting the judgment of scientists; the fact that a strong and outraged denial of fraud by the perpetrator often stops people from pursuing the matter; and even – and some have suggested that this might turn out to have been a factor in Heslop Harrison's case – the desire of the student to achieve results that will please the master.

Both cases can be described in terms that have a familiar ring to them, now that we have followed the Heslop Harrison story so far:

Case I. A scientist believes strongly in a particular theory. Many other scientists in the same field do not share this belief. The scientist produces data he has obtained that confirm the theory. Other scientists try to produce the same data but fail. Accusations of fraud are heard at informal gatherings of scientists. When doubts are raised about the scientist's results, he says that his colleagues are just not good enough to produce the same data, but that the data are there. The scientist dies and the accusations, though still believed by many, are never published or pursued.

Many years ago I investigated a field called biofeedback for a television documentary and came across the work of Neal Miller, an eminent psychology professor at Rockefeller University in New York. His lab had been carrying out pioneering research into what was called operant conditioning of autonomic responses. Our nervous systems have two divisions, one under voluntary control and the other, the autonomic nervous system, controlling such processes as digestion, heart rate, blood pressure, kidney function, and so on, entirely beyond our voluntary control. Or, at least, that's what people thought until Miller and others in the field began to carry out experiments that contradicted the conventional view.

It is easy to see the benefits of an autonomic nervous system. Our bodies have enough to think about, from minute to minute, without having to decide consciously what heart rate to use, or whether it's time to move today's lunch a little further down the large intestine. However, some conscious control of such involuntary functions might be useful. For example, if people with high blood pressure could learn to reduce their blood pressure voluntarily, without the use of drugs, this would be of great benefit.

Miller and his colleagues were looking into whether it was possible to condition rats to lower their blood pressure or change their heart rate, believing that it was. The early data were impressive. The rats in Miller's lab *could* be conditioned to produce significant changes in their own heart rates by being rewarded for a positive result, and the news encouraged other scientists to research different types of autonomic conditioning and even to try out possible treatments for high blood pressure or irregular heart rate, using what was called biofeedback.

Twenty-five years after I first looked into the topic, I was visiting Edward Katkin, a senior scholar in the field of psychophysiology. Katkin has a sharp analytical mind and is an acute observer of the doings of his fellow scientists. I described to Katkin my interest in scientific fraud, and he unfolded the story of Neal Miller's lab and its successful demonstration of biofeedback for a number of physiological functions that were usually beyond voluntary control.

In Miller's lab, the research was carried out by several researchers working under him, and some of the most impressive results were obtained by a postdoctoral fellow named Leo DiCara. 'Among the most influential of Miller's papers,' Katkin said, 'were the ones that were published with Leo DiCara. They were unusually sophisticated and used a variety of dependent measures that defied the arguments of critics. DiCara and Miller demonstrated that they could condition a rat either to decrease or increase its heart rate; they demonstrated conditioning of filtration rates of the kidney, and they trained rats to shift their blood from one side of the body to the other. These dramatic demonstrations resulted, according to generally accepted wisdom, in Miller's getting onto the shortlist for the Nobel Prize. All of this was documented well in a series of *New Yorker* articles by Gerald Jonas. Miller's papers also led directly to the establishment of an entire new discipline, biofeedback, and the establishment of new research societies focusing on the impact of learning theory and behaviourism on medicine. These developments continue to flourish. Gradually and insidiously, however, rumours began to circulate among the "insiders" that something was going wrong in Miller's lab. Some of us had heard that DiCara's work was unreplicable and that Miller was beside himself with distress.'

Jasper Brener was a South African scientist who worked on nervous-system conditioning and who became involved in investigating some of the most dramatic biofeedback results produced by Neal Miller's lab. He described his research to me in his office at the State University of New York at Stony Brook, where he worked alongside Ed Katkin in the Department of Psychology. I could see him very much in the John Raven mould at the time of the events he was describing – passionate about his subject and affronted by the idea that someone would betray truth in the way he believed DiCara had. 'All the people I knew who were involved in this accepted that Leo fudged the data,' he said. A scientist who had worked in DiCara's lab even did the rounds of conferences and meetings, telling colleagues involved in the research that DiCara's work was faked. But again, as with Heslop Harrison, no one wanted to publish specific accusations. In this case, the motive for reticence was the desire not to offend or embarrass Neal Miller, who was already believed by some to have suffered enough from the effects of the failure to replicate DiCara's work.

Brener had tried to replicate DiCara's results and his lack of success reinforced the growing belief that something was wrong either with DiCara's description of his experimental method, or with the data he published, or both. 'At the time,' said Brener, 'I was so deeply immersed in this, it became a crusade to show that he must have fudged his data.' That crusade sounded not very different from Raven's, but, unlike Raven's, Brener's view of his antagonist was tinged with warmth: 'He was a regular guy, pretty flamboyant. I liked him.'

People often don't understand the importance of replication in science. It is sometimes seen as too stringent a condition by poor scientists who have obtained a result that is difficult to repeat. But you do have to repeat it – or other people do – if your colleagues are to be convinced, because there are all sorts of reasons something might appear to occur once and never again, ranging from poor observation to faulty apparatus. So if you reproduce exactly the conditions and events that led to a result on one day, it should happen on any other day. If it doesn't, there's something to worry about. Brener told me of a revealing remark DiCara once made to him: 'If you see it once, it happens.' Of course, if you believe that, there's no need to repeat the result to convince yourself. And if it turns out to be more difficult to repeat your result than you thought, as with Heslop Harrison, 'fudging' your data may seem only a venial sin if the overall effect is a good one – to persuade your colleagues of the truth of a theory you already know is true. 'If you see it once, it happens.'

When Brener tried to replicate DiCara's results, he failed. DiCara had kept his rats on an artificial ventilator as a necessary part of the experiment, and Brener found, when trying to reproduce the effect, that by altering the way in which the rats were artificially respirated, he could change the heart-rate and blood-pressure responses in the paralysed rats, an effect that had nothing to do with the rats' own abilities, but was produced entirely by the actions of the experimenter.

What Brener believes happened – although this is sheer speculation – is that DiCara's description of the artificial respiration method in the report of his results was wrong and that DiCara probably changed the breathing rates and pressures during the experiments, and this produced the changes in heart rate and blood pressure that he said were due to conditioning.

At about the time Brener and others were failing to replicate DiCara's results, several things happened in quick succession. First DiCara, who had been born and raised in New York and hated leaving the city, took a job at a university in Michigan, a long way from the buzz and excitement of his home. Next, the failures to replicate were reported at scientific meetings, where he was hounded by the other scientists, to whom he said, 'You didn't do the experiments right.'

Neal Miller was scheduled to present the latest data on his groundbreaking work in April 1972, at a major conference in Philadelphia. Katkin was at that meeting and remembers it vividly. In 1998 he told me, 'The largest meeting room at the convention was set aside for [Miller's] talk, and the room was packed to overflowing, with loudspeakers placed outside so the overflow crowd could hear him. Then Miller dropped the bombshell that until that day had been known only to the "insiders". He had been unable to replicate his own great successes. He presented a graph on which one axis was the years from 1966 to the present and the other was the number of successful demonstrations of autonomic conditioning in his laboratory. The graph was a beautiful decelerating curve. Miller never mentioned any names, but sophisticated observers could see that the rate of success dropped off in DiCara's last year in the lab and fell to zero after his departure. I remember a number of my friends discussing this over dinner that night, and we all concurred that the systematic variable was the dreaded DiCara. Miller never would even hint at that, and I do not know anybody who had the courage to discuss it with him directly. Certainly I never did.'

From that day on, DiCara suffered terribly. At subsequent meetings, symposia were held to try to determine what went wrong. DiCara would participate and be subjected to withering examination. He was in obvious psychological pain, asserting repeatedly that nobody knew the proper technique and that as soon as his new lab in Michigan was established he would invite all to come and watch him replicate the findings. The new lab in Michigan never came to be. After moving to Michigan, DiCara killed himself.

None of these events in itself proves anything about DiCara's research. DiCara's unexpected move to a faraway state doesn't prove that he was asked to leave by Miller and sent to a prearranged job a thousand miles away, as some have suggested. (Miller has never spoken publicly about the affair.) The failure to replicate doesn't prove that the original work was faked. Biological experiments are complex and time-consuming to set up, and it is sometimes difficult to describe the process in such detail that another scientist could replicate it exactly, particularly when you don't actually know which details are crucial to producing the results. And DiCara's suicide tells us nothing about the origins of the state of mind that led him to end his life.

But as with Heslop Harrison, the story was accompanied by a strong and almost universal belief among people in the field that DiCara had faked his results, and nobody did anything about it.

The next story is an example of how, taken to extremes, the erroneous or fraudulent reporting of what may be quite insignificant discoveries can pollute the well of scientific discourse in a field. It helps us understand why we should treat scientific fraud, including the allegations made against Heslop Harrison, seriously.

Case 2. Over a period of twenty-five years a scientist who is particularly familiar with a certain geographical region reports a series of discoveries he has made in the area. These discoveries surprise the other scientists in his field for two reasons: they support a theory about the history of the area that few scientists believe; and discoveries like these have been reported previously only in other parts of the world. But in many scientific papers, often co-authored with other – reputable – scientists, the scientist builds up a picture that, if correct, overturns the conclusions derived from much of the major evidence uncovered by other scientists about this area. One day, a scientist who has become suspicious of aspects of this research begins an investigation that results in a paper in *Nature* claiming that the so-called discoveries had actually been introduced to the area from outside. But the only mention of the suspect scientist is an oblique reference in a footnote. It is clear to everyone in the field at whom the finger is pointed, but the scientist in question denies absolutely that any fraud has taken place.

This story concerns palaeontology, and the investigation began when an Australian palaeontologist, John Talent, was in Paris in August 1986. He visited a well-known fossil and rock shop and bought an ammonoid from Erfoud in Morocco. For some years,

Talent had been puzzled by a series of articles by an Indian palae-ontologist, Viswa Jit Gupta. They reported a range of fossilised life forms, from ammonoids to conodonts, plants to primates, all alleg-edly discovered in the Himalayas, and all unexpected. It was a year or so after Talent bought the ammonoid that he noticed in a paper by Gupta an almost identical specimen illustrated – improbably – as a Himalayan find. The authority of the paper was strengthened because a co-author was a well-known and respected expert on this particular type of ammonoid. As Talent delved into Gupta's other publications, he noticed that this was a pattern: improbable or unprecedented observations were accompanied by the pairing of Gupta's name with that of a distinguished scientist in the field. These scientists would supply comments and analysis of the particu-lar fossil, but in no case had they been present at the finding of the fossil, or even in India.

One significant series of papers reported on conodonts, tiny tooth-like fossils believed to be part of the jaw apparatus of primi-tive worms or other creatures, and described examples that were supposedly found from one end of the Himalayas to the other, at various levels widely separated in period of origin. This was surpris-ing enough – that a very distinct type of fossil should be so widely spread in time and space. What was even more unusual was that, until the point when Gupta described these new finds, such fossils had only ever been known in one particular part of the world: the North Evans limestone at Amsdell Creek, New York. Now, mirac-ulously, identical fossils had turned up in the Himalayas, but they had never been seen anywhere in between. Furthermore, it so happened that there were plenty of samples of conodonts around, if anybody had wanted to get hold of some and 'plant' them some-where as inappropriate as the Himalayas. At the time this fossil bed was discovered – a particularly rich and peculiar collection of objects – specimens were distributed freely to other conodont workers and to many schools and universities.

When Talent looked in more detail at some of the conodont papers, he discovered another very suspicious fact. In one paper, Gupta showed a photograph of one of these fossils and used it to establish a date for a particular geological formation in the Himalayas. (If the date of a particular fossil type is known, its presence in a geological layer suggests a similar date for the rock in that layer.) Then, in another paper, Talent discovered an identical photograph of the same fossil being used to establish the same date for an entirely different geographical location.

Talent had been fortunate enough to gain entry during the early 1970s to one of the areas of the Himalayas that Gupta had studied. He described his observations to the American periodical *Science*:[41]

> The first thing we did on the ground just happened to follow up something Gupta had published on. The trouble was, nothing matched up. In 1966 Gupta had reported finding graptolites* from this locality, in a paper with Bill Berry at Berkeley. When we looked, not only did we infer quite a different age for the rocks, but it was clear that no fossils could have come from there. The rocks were intensely deformed, so that no fossils were likely to have survived, least of all the fragile little graptolites.

From the beginning, according to Talent, the Gupta pattern was set: 'suspect fossils, spurious age determinations, doubts about the localities'. Just as with Heslop Harrison, Gupta often failed to make clear exactly where he had found various specimens. 'Gupta's descriptions of localities are often extremely vague,' Talent told *Science*, 'so vague as to make relocation impossible.' One of these locality specifications, said to be the source of some Jurassic ammonites, was a 130-kilometre stretch of a Himalayan valley wall, a

---

* Small extinct marine animals.

degree of imprecision that virtually precludes relocation by other scientists.

Encouraged on one occasion to be more informative about a graptolite locality, Gupta explained that he was deliberately being vague because other geologists were working in the area and might try to pre-empt his publication if they found the fossil site. (Shades of Heslop Harrison and his protectiveness of his Hebridean territory.)

The roll call of Gupta's suspicious findings is long and complex. In a paper in *Nature*, Talent wrote:

> Over the past 20 years the Himalayas have been the scene of many startling discoveries involving species previously known in great abundance from specific places on other continents, that can be said to be 'well fingerprinted' by highly characteristic modes of preservation, and that had long been in teaching collections around the world.

Talent listed some of the most dramatic examples, then wrote,

> There are a host of similar examples from the Himalayas that are paleobiogeographically enigmatic. If correct, these reports would indicate that the global distributional data accumulated over the past century and synthesized into paleobiogeographical patterns has been seriously – even wildly – out of kilter for most of the Palaeozoic and Mesozoic.[42]

Even after the compelling evidence suggesting that deliberate malpractice was at the roots of the events he described, Talent at no point said 'I maintain that Gupta collected fossils from shops, muse-ums, and universities around the world and "planted" them in the Himalayas.' In fact, just as Raven's *Nature* letter seemed to be a neutral identification of certain plants as aliens, Talent's *Nature* article seemed to be a plea for scientists to set out and verify Gupta's huge number of dubious reports. That this was not a serious suggestion is clear from the fact, reported in *Nature* and in other articles, that most of the areas where the discoveries were made are in parts of the Himalayas that, for security reasons, are barred to foreigners.

Talent first went public with his discoveries at a geological congress in Calgary in August 1987, a year after his visit to Paris. 'Many people in the field learned of this business at the 1987 Calgary meeting,' he told me, 'but there were many others who didn't. Gupta continued in the same way as before, and more researchers became involved in useless papers with him.'[43]

The situation produced what journalists like to call red faces among many of the scientists who had been taken in by Gupta. 'My face is red, no question about that,' said Gary Webster, of Washington State University, co-author of nine papers with the Indian palaeontologist. 'In the early seventies I asked to see a specimen of species of crinoid* he'd published on; it looked interesting, because it was the first crinoid of this species reported outside the United States. He kept sending me more material, and we produced this series of papers. Then I heard Talent's talk at the Calgary meeting. I was aghast. I am now virtually certain that most of these specimens did come from places other than the Himalayas. I certainly should have been more wary.'

Other scientists came forward after Talent went public with the story and provided corroborative evidence of their own. One of them was Philippe Janvier, of the Museum of Natural History in Paris. He told *Science*:

Gupta was visiting me in Paris, after a trip to China. He told me he had a magnificent fossil fish skull, and showed it to me. I

---

* Marine invertebrates also known as sea lilies.

could see it was a new species, and agreed to write a paper with him. Shortly after that I went to Sweden on a trip, and visited the palaeontological museum, where I met Zhang Miman, who is director of the Institute of Palaeontology in Beijing. Zhang was working on some fish fossils which she had brought with her from China, and I immediately recognised a specimen of the same species that Gupta had showed me – same colour, same matrix, everything. I was shocked, but didn't say anything immediately. I returned a few weeks later and asked Zhang about the fish. She told me that they were relatively common in China, and they were frequently offered to visitors as gifts. I immediately telegraphed Gupta, telling him to drop the new species name for the fish. Now, there is no evidence that Gupta brought the fish fossil with him from China, but I'm 99 per cent sure he did. Many people have been had by Gupta in this way.[44]

Other scientists interviewed by *Science* ruminated on some of the reasons that allow fraudulent behaviour to occur in science.

Walter Sweet, of Ohio State University, said, 'If a colleague sends you a specimen and says it comes from such and such locality, you don't immediately suspect it might not be true.' Art Boucot, of Oregon State University, explained, 'We are not trained to think in terms of dishonesty in science – it's an absolute no-no. We are pretty innocent in this regard.' And a *Nature* editorial in the same issue as Talent's article said, '. . . the only durable remedy is to create an intellectual climate in which people's willingness to hazard colleagues' trust is diminished to zero.'

The scientist at the centre of the story, Viswa Gupta, responded to the *Nature* article with a counter-charge worthy of Heslop Harrison under attack by Raven. He managed to get his personal remarks about Talent reported by *Nature* a week later. According to Gupta, Talent's attack was motivated by 'a malicious intent to take revenge for personal rivalry and professional jealousy over the past rs.' Gupta went on to say that his twenty-five years of work had documented 'the existence of Silurian-Devonian succession in the Himalayas' (a geological theory he had been trying to prove) and that 'the truth about it will be proved by the progeny during the forthcoming years'.[45]

Talent concluded his *Nature* article as follows, starting with a paragraph that might have been transposed wholesale as a comment on the Heslop Harrison case:

> Most scientists are reluctant to 'blow the whistle' on a colleague; and most tend to have highly specialised research interests. So even if a specific item of data is suspected of being false, it would directly affect very few people. The tendency is to dismiss such episodes as the one-off aberrations of scientists straying outside their accustomed bailiwicks, rather than as elements of a potentially much larger fabric: in the present case an especially vast and intricate fabric.
>
> Rhinos in Rio? Kangaroos in Kashmir? Well, something as remarkable biogeographically is said to have occurred. At first sight it might appear that a whole circus of exotica – mainly invertebrate – was let loose and fossilised seriatim in the Palaeozoic and Mesozoic sequences of the Himalayas. Earth scientists in general, and palaeontologists in particular, have blissfully assumed that, apart from the Piltdown Man, their science was largely free from attempts to pollute the literature. There have been cases of practical jokes, and examples of misappropriation of materials by individuals overeager to publish. But compared with the cornucopia of items disgorged into the stratigraphy of the Himalayas region over the past 25 years, such instances are mere bagatelles.

These two stories – from psychology and palaeontology – are among dozens of examples of alleged fraud I've come across in the literature and, I suspect, dozens more that haven't strayed outside the bar conversations of scientists. They tell us nothing directly about the Heslop Harrison story, but they are loaded with similarities – of possible motive and method, for a start, but also in the way other scientists reacted to the allegations, pursued them, passed them on, and judged the accused, and in the way scientists accused of fraud have responded when faced with the allegations. Since none has been proven in a court of law, they *could* simply demonstrate that scientists are continually accusing their innocent colleagues of fraud – or, as I believe, that there will always be scientists who use fraud to reinforce support for their theories.

And in that sense they tell us something indirectly about the Heslop Harrison case: that such things happen, frequently and everywhere. Whatever the facts about events in the Hebrides in the 1930s and 1940s, no one can dismiss Raven's case against Heslop Harrison because of its implausibility. As the stories of DiCara and Gupta, and many, many others show, it is all too plausible.

And the reasons for such behaviour have been well summarised by Felix Franks in his book *Polywater*, which explored a story not of fakery but of another type of deviant science, an example of self-deception that took place in the 1960s. In analysing that story, Franks wrote:

> If ever the delicate balance between making a living, searching for the truth, and obtaining the approbation of one's peers is upset, then there is the danger of deviant science. The signs are familiar to most professional scientists: the race to publish gathers momentum, speculation takes the place of information, competitors are accused of plagiarism or selective use of evidence, and, commonest of all, publications take on a partisan character (they might for instance omit any mention of related work performed by others). This intense competition evades and sometimes

violates the accepted professional norms . . . There is [an] incentive to dishonesty among scholars. Scientists can take up, or be forced into, positions in defence of theories that make it hard, even impossible, to extricate themselves 'with honour' once the weight of evidence mounts up against them. In such situations there have been cases of forged experimental results.[46]

I will end this chapter with a fictional insight from C.P. Snow's novel of Cambridge life, *The Affair*, which might help clarify the picture – or preserve it in ambiguity. It is an attempt by the protagonist, Lewis Eliot, to suggest to a committee of fellows of the college why a distinguished scientist might fake evidence:

'You're asking us to believe,' [says Nightingale, one of the fellows], 'that a man absolutely established, right at the top of his particular tree, is going to commit forgery just for the sake of that?'

'I did my best,' [says Eliot, referring to earlier cases of scientific fraud]. 'These frauds had happened. We knew nothing, or almost nothing, about the motives. In no case did money come in – in one, conceivably the crude desire to get a job. The rest were quite mysterious. If one had known any of the men intimately would one have understood?'

Anyone's guess was as good as mine. But it didn't seem impossible to imagine what might have led some of them on, especially the more distinguished, those in positions comparable with Palairet's [the fraudster]. Wasn't one of the motives a curious kind of vanity? 'I have been right so often. I know I'm right this time. This is the way the world was designed. If the evidence isn't forthcoming, then just for the present I'll produce the evidence. It will show everyone that I am right. Then no doubt in the future, others will do experiments and show how right I was.' . . . Couldn't one at least imagine him getting old and impatient, knowing he hadn't much time, working on his last problem, not an important one, if you like, but one he was certain he knew the answer to? Certain that he knew how the world was designed? Almost as though it was the world designed by him. And mixed with that, perhaps, a spirit of mischief, such as one sometimes finds in the vain-and-modest – 'this is what I can get away with.'

11

'A Total Muddlehead'

Raven had been told that some of Heslop Harrison's Newcastle colleagues wanted to be 'in at the kill' but in the event there was no 'kill'. The time to act in any sudden, forceful way would have been when Raven's letter was published in *Nature*, but instead there was silence, although there was an undercurrent of behind-the-scenes activity on and off for the next twenty years.

The important thing, it seemed to the botanists, was that dubious attributions should eventually be expunged, whether or not Heslop Harrison himself was incriminated. The views of people like Vero Wynne-Edwards prevailed. 'No one wants such things to happen,' he had written to Raven, taking upon his shoulders – somewhat presumptuously, I feel – the responsibility of speaking for the entire profession of botany.

Whether or not everyone agreed that the public disgrace and humiliation of Heslop Harrison would be a bad thing, most people accepted the evidence that he had done wrong. There were several reasons for this. Of course, Raven's report, known in outline if not in detail to many, provided evidence that wrongdoing had occurred. But two other factors strengthened people's belief. One was that Heslop Harrison had a motive for such wrongdoing: strengthening the evidence for the periglacial theory. The other was that there were rumours of research activities by Heslop Harrison in other fields that also seemed suspicious.

I discovered this one day when I was discussing Heslop Harrison's motives with Dr 'O'Connor', the botanist who preferred to remain anonymous. He dropped a small bombshell into the conversation: 'Of course, Heslop Harrison had been suspected of similar behaviour over moths, you know, some years before.' Rummaging in his bookshelves, O'Connor produced a book called *Ecological Genetics*,[47] by E.B. Ford, and we were off on a new trail.

Edmund Brisco Ford, known as Henry to his friends, was a leading figure in the world of evolutionary biology in the first half of this century, as well as being an ardent collector of moths and butterflies. He was a rigorous and intelligent scientist, a careful experimenter, and according to one scientist I spoke to who knew him, 'an absolute shit'. His shittiness lay partly in his misogyny, illustrated in one often-repeated story: he arrived once to give a lecture, found only women waiting to hear him and said, 'How extraordinary – there is no one here.'[48] He was also accused of snobbery and rudeness to people he did not like. Clearly, as far as the last two characteristics were concerned, he was a man fated to clash with Heslop Harrison, the ironworker's son who was very rude to people *he* did not like. In his book, Ford described Heslop Harrison's research on two species of insect, and heavily criticised the results and their interpretation.

In fact, in the way that occurs with antagonists, there were certain similarities between them. I noticed on reading one of Heslop Harrison's papers[49] that he had the habit of referring to people he was quoting as 'my friend': 'My old friend T.A. Lofthouse', 'my old friend Lange', even 'my friend Kettlewell', a scientist who was actually one of the strongest critics of Heslop Harrison's work on moths. Then, in E.B. Ford's Royal Society obituary I read: 'Other biologists were designated as "my friend Theodosius Dobzhansky" (relishing every syllable), "my friend Sir Ronald Fisher", and so on. Once he even produced "my friend the Pope."'

The background to Ford's quarrel with Heslop Harrison lay in a disagreement over the way in which inheritance works, and in particular, in a process called the inheritance of acquired characteristics. Heslop Harrison believed that this process did occur and that it played an important part in heredity. Furthermore, he carried out two different sets of experiments that provided proof of his ideas.

Every schoolboy knows about melanism in moths. It is one of those examples brought up in biology lessons that show how Darwinian evolution – evolution by natural selection – works. Nowadays, the story is told in the following way. In the north of England, in the days of clean air, a particular species of moth displayed a variety of wing shades among its individuals, ranging from light to dark, but most of the moths had light-coloured wings. As heavy industry developed in the area, it polluted the environment, coating trees and leaves with smoke residue and noxious chemicals. Entomologists then noticed that more and more moths of this species had dark wings, and fewer and fewer had light ones. This process, called melanism after the Greek word for 'black', demanded an explanation and it got two. One of these explanations was favoured, indeed invented, by Heslop Harrison. The other, now generally taught in schools, is that evolution by natural selection had occurred.

Evolution by natural selection is often just called evolution. But, in fact, this misses the most important bit. Few people would or could deny that evolution occurs. There are observable changes in successive generations of living creatures so that characteristics of an earlier generation disappear from later generations, and others appear that weren't there before. These changes often accompany alterations in the environment and make it easier for the organism to survive in the new environment. This process of change could be explained in a variety of ways, and it wasn't until the nineteenth century that a plausible scientific explanation – evolution by natural selection – was offered by Charles Darwin.

Darwin's great theory offers an explanation for many of the facts of biology that had puzzled scientists and natural historians for centuries. How did different species arise on the earth if they weren't created separately by a divine Creator? The fossil record that was fleshed out in the eighteenth and nineteenth centuries showed that creatures used to exist that were very different from current models. It was likely that successive generations of breeding had led in a line from those fossil animals to modern creatures, and that several different species today were derived in some way from a single species millions of years ago.

Darwin's theory, at which he arrived after decades of careful observation of species around the world, was that each successive generation of a living creature – and the same applied to plants – inherited characteristics from its parents. Which characteristics were inherited from which parent was determined by chance, so that a blue-eyed (or black-furred) parent might give birth to offspring whose eyes (or fur) might be blue (or black), but she might not, if the offspring had the eye or fur colour of its father, or even of some more distant relative.

These differences that occurred by chance, from one generation to the next, were a fertile source of adaptation – the process by which future generations became more suited, or 'fitted', to their environment, better able to cope with hazards of geography, climate, or predators (hence, survival of the 'fittest'). In Darwin's view, the process worked solely through competition. In any generation of organisms, the random allocation of new characteristics would lead to some organisms that were better off than their peers and others that were worse off, in terms of how well their collected adaptations enabled them to cope with life. So the less well adapted died sooner or in greater numbers and had fewer offspring, while the better-adapted animals or plants survived longer and in greater numbers, had more offspring, and came to predominate. It was such adaptations that eventually led to new species as, over many generations in a changing environment, the individual characteristics changed the organism so much from its earlier form that the two types could no longer breed.

By the early part of the twentieth century, when Heslop Harrison was growing up and beginning his study of natural history, Darwinian theory was fairly well understood as a tool for interpreting the relationships between different members of the animal and vegetable kingdoms. What was not understood was the mechanism by which the information about individual characteristics was transmitted from one generation to the next, and how the changes occurred that produced novel combinations of characteristics in the next generation. Because of that lack of understanding, it was possible to entertain the idea of various mechanisms of transmission. Heslop Harrison firmly believed in a theory that most scientists would now say is erroneous, and he even found evidence to prove it.

The relevance of evolutionary theory to botany is that it provided an understanding of how plants bred. Indeed, in addition to Darwin, Gregor Mendel, the Czech monk who studied pea plants, provided the basis for understanding what would happen if one plant breed combined with another to produce a third. It is a small irony of this story that Mendel is another scientist whose research, providing results which so neatly supported the theory he held, has since come under suspicion of fakery, or at least of vigorous massaging of the evidence.[50]

So back to the moths. The explanation of melanism in moths supposes that in a particular generation of moths those that contrasted most with the background would be picked off and eaten by birds. If the trees and vegetation were lighter, it would be the dark-winged moths that would be eaten first, leaving lighter-winged insects to breed descendants that, on the whole, would also have light wings. But as the trees and leaves got darker with the industrial pollution, the darker-winged creatures, formerly in the minority, would be less visible than the lighter ones, so they would survive longer to breed more descendants. In this way, an environmental change led to a genetic change, not in the individual but in the population. No single moth changed its genes as a result of pollution, but later generations of moths had different genes for wing colour than did earlier generations.

It is tempting to say that the change of colour occurred 'in order to' conceal the darker moths from predators more effectively, but it is not really 'in order to' at all. One of the extraordinary things about Darwin's theory of evolution, and also the most difficult to grasp, is the way in which it leads to an appearance of cause and effect that does not exist, at least for the individual members of the species. To take a more familiar example, the giraffe did not develop a longer neck 'in order to' reach the topmost branches of a tall tree. Those giraffes that were born with slightly longer necks in an environment where the trees had tall trunks ate more leaves than their shorter cousins, survived longer, and had more long-necked children. Nowhere in that chain of events is there room for an intention, an 'in order to'.

Before Darwin – and continuing after him – came a range of theories in which intention did play a part. Formulated by Jean-Baptiste Lamarck in the eighteenth century, the theory of the inheritance of acquired characteristics provided a convincing explanation for how species evolved. Here, the giraffe's long neck really does result from an 'intention' to reach the higher branches. Lamarck, and later supporters of his ideas, thought that an individual giraffe, by striving to reach the higher branches, could somehow add an imperceptible fraction of a cubit to his height and that this increase could somehow be passed on to his children. Equipped with that tiny extra bit of height, the child giraffe would grow up, do his own bit of striving, and add a further sliver or so. Over generations, the process would lead to dramatically longer necks.

There is one factor to note in all this. In the case of the giraffe, the lengthening process is dependent on a change in the environment: the lack of foliage at levels at which the shorter-necked ancestors of the giraffe could eat it and survive.

By this century, Lamarckian explanations had become rather more sophisticated than the idea of a creature's striving to change to fit into a changing environment. But the principle was the same: some physical change that an individual member of a species underwent during its life could be transmitted in a quite specific way to the next generation.

One of Britain's leading evolutionary geneticists was John Maynard Smith, a professor at the University of Sussex and a colourful, outspoken advocate of Darwinian evolutionary theory. In his youth, Maynard Smith was a Marxist, and he described to me how passionately he wanted to believe in Lamarckian ideas, because they would confirm what all good Marxists believed: that a firm basis of good education would lead to the widespread improvement of future generations of citizens, as they inherited the improved skills and intelligence of their forebears. 'I tried very hard with some experiments of my own to prove Lamarckism and was bitterly disappointed when I failed,' he said. But Maynard Smith, unlike Heslop Harrison, overcame his disappointment at being unable to provide evidence for the theory and became a leading supporter and analyst of Darwinian ideas, now generally accepted by almost all evolutionary scientists in some form or other, with dissent confined to detailed issues of genetic mechanisms, mutation rates, and the intricacies of organ formation.

Maynard Smith told me that, when he was a graduate student in the 1950s, he heard vague rumours of Heslop Harrison's alleged 'plantings'. 'Weren't they discovered in John Innes Potting Compound No. 2 or something similar?' he said, showing how rumour and exaggeration had worked on Raven's admixture of quartz among the olivine soil of Rum. Maynard Smith did not have much time for biologists of the Heslop Harrison sort. 'Many people with chairs of biology in those days never understood melanism – they were Lamarckists without thinking about it. I remember [Sir Peter] Medawar saying, "Palaeontology is a particularly undemanding branch of science," and there's a sense in which biology in those days was an intellectually undemanding science too.'

The fact is, however, that in the 1930s there was quite a strong Lamarckian tendency among some scientists; it was also easier to believe in these ideas before a detailed understanding of genetics showed why they were highly unlikely to be true. A letter in *Nature* in 1932 from Professor E.W. McBride replied to criticism of a lecture he had given on the topic at the Royal Institution, quoting some of Heslop Harrison's experiments.

> Professor [J.B.S.] Haldane directly challenges them [Heslop Harrison's findings], which seems a rash proceeding, considering that Prof. Harrison occupies a chair of botany and is unrivalled for his knowledge of the systematics and ecology of the British flora . . . As I gave a list of eminent authorities who had become converted to Lamarckism, Prof. Haldane quotes against me the Royal Society motto, *Nullius in verba** . . . But the opinion of an expert in systematics is not to be dismissed as mere '*verba*'. It is really a deduction from thousands of relevant facts which the critic has neither the time nor experience to be able to consider.

It is always difficult to know why people find one scientific theory more convincing than another, since each argument is usually persuasive to some group of experts. It might have been comforting for Heslop Harrison, as a self-made, successful, upwardly mobile son of the working classes, to know that his descendants would start off life with the benefits of his success. On

---

* Usually translated as 'Take nobody's word for it'.

the other hand, it might have been an easier and less intellectually demanding explanation of how evolution works, with an intuitive charm and without the problems presented in Darwinian theory by time scale, sophisticated organ formation, and no apparent mechanism of effect. But for whatever reason, Lamarckian ideas were at the heart of Heslop Harrison's entomological work, in two series of experiments: on melanism in moths, and on the food preferences of sawfly larvae.

Until his dying day, Heslop Harrison believed that the melanism of moths was due to some kind of pollution of the individual, which was passed on to its descendants. In other words, he believed that the explanation did not lie in smoke-darkened tree trunks aiding camouflage in the darker moths, but in some effect of chemical pollutants on the metabolism of the moth and on the genes it passed on to its descendants. Characteristically for Heslop Harrison, he began his explanation with an implied attack on the darkened tree trunk theory by offering an unsupported personal observation that the reader is required to accept: 'Despite the truth of my observations that tree-trunks are not significantly darkened, every leaf bears a film of smoke-derived impurities composed of organic compounds and various salts of . . . metals.'[51]

He went on to set out a theory by which chemicals eaten by the moth somehow found their way into just that part of the chromosome that controlled the moth's ability to survive the effects of those chemicals. John Maynard Smith quoted another biologist who described the inheritance of acquired characteristics as being 'as if a man sent a telegram in English to China and it arrived of its own accord translated into Chinese'.

This is always the problem with Lamarckism. A particular characteristic – say, wing colour – is determined by an arrangement of atoms along the spiral molecule DNA, which carries the instructions for the developing larva. Even if the environmental influence – a noxious chemical, in the case of the moth – found its way to the egg from which the next moth develops, it is most unlikely to produce any meaningful change in the moth's genetics. All it would do is change a few atoms here and there at random, probably destroy the information the gene carries, and produce nonsense rather than becoming a different set of instructions – to produce pigment rather than no pigment, for example. It is as if a sentence in a recipe for scones that says 'Add salt and cheese' had had some letters changed randomly so that it then says, miraculously, 'Add sugar and raisins', producing a different but equally delicious type of scone. It would be far more likely to read, 'Adg selb tnd lhuece' and leave the cook thoroughly confused.

Nevertheless, Heslop Harrison was determined to prove that his theory was correct, motivated, it seemed, as much by his desire to poke holes in the accepted Darwinian theory as by the plausibility of the Lamarckian explanation. As seemed to happen quite often with Heslop Harrison, he rode off on his hobbyhorse in search of evidence to prove his theory and found it.

Into kitchen, bathroom, or tool shed, Heslop Harrison took some larvae of a particular moth not known for its melanistic tendencies and fed them on a diet of foliage impregnated with some of the nastiest by-products of industrial pollution: salts of manganese and lead. After a number of successive generations were treated in this way, without – apparently – being poisoned in their cradles by this noxious food, Heslop Harrison claimed to find that dark-winged individuals had appeared and that further breeding without the chemical feed led to dark-winged offspring.

In his book *Ecological Genetics*,[52] E.B. Ford summarised the experiments and dismissed Heslop Harrison's results as due to a mixture of poor methodology and bad scientific reasoning: 'Fisher showed that Harrison's controls were not adequate . . . Harrison seems never to have considered the mutation-rate involved . . . McKenny Hughes as well as Thomsen and Lemche have repeated Harrison's experiments . . . they were entirely unable to confirm

his results; no melanics at all appeared in their treated stocks . . .'
And later: 'An attempt made by Harrison to study the predation of
resting moths failed, from which he reached the unjustified conclusion that birds are of negligible importance in this connection . . .'

Comments by other scientists on this research were universally
negative, even if carefully phrased, but, says A.D. Peacock, in
Heslop Harrison's Royal Society obituary, 'To these serious criticisms Harrison made no comment but held resolutely to his
opinion.'

No one had obtained such an amazing result before. No one has
obtained it since. Had the evidence been valid and the explanation
correct, Heslop Harrison would have achieved a small, or even
medium-size, triumph of genetics. But, as Peacock observes, the
'validity of Harrison's results has been challenged for various
reasons. Certain workers failed to confirm his experimental findings; but Harrison (1935) attributed their failures to faults in techniques employed and dismissed their criticisms.' Heslop Harrison
showed not a scintilla of doubt about the correctness of his ideas.

The melanism story is one example of a 'finding' by Heslop
Harrison in entomology that didn't hold up. His related studies on
sawflies were also never repeated.[53] Here, he carried out a series of
experiments with insects that used for food a particular species of
willow and no other. Heslop Harrison believed that this food preference was created when the sawfly laid its eggs on the willow and
the larvae developed on the leaves. He further believed that this
preference was passed on genetically to its offspring when the
sawfly became an adult. If this was truly a Lamarckian event, Heslop
Harrison hoped to induce a preference for willow type B in all the
descendants by forcing the larvae of a species that preferred willow
type A to feed on willow type B. This hope was fulfilled, and, as
Heslop Harrison said at the end of his paper reporting the work, 'A
direct demonstration of a genuine Lamarckian effect has been
made, and one is forced to admit that Lamarckism itself has received

very powerful support, both directly and indirectly, from the work described above.' Once again, those key Heslop Harrison words – 'forced to admit' – make their appearance.

In fact, many scientists were not forced to admit anything of the sort, and over the next thirty years Heslop Harrison's moth and sawfly work, regularly referred to in his later papers, was under constant fire from scientists who tried and failed to replicate it or who came up with alternative, non-Lamarckian, explanations for the observations.

I asked John Maynard Smith what he thought Heslop Harrison had actually done all those years ago. Was this another example of fraud, as some scientists believed? Maynard Smith was more charitable, sort of. 'No, I don't think they were fakes – the fellow was just a total muddlehead.'

The significance of this work for the broader story of fraud allegations against Heslop Harrison is that it contributed to a climate of disbelief that came to surround all of his researches. Somehow this man was too eager to prove things, by hook or by crook, and there were plenty who believed that 'crook' was the word to use.

What I found surprising about Heslop Harrison's Royal Society obituary was how it revealed that many, if not most, of the grand ideas he presented in his papers over the years were eventually discredited. At the time of his researches, when less was known about the detailed mechanisms of genetics and geology, it might charitably be said that his ideas were actually as valid as anyone else's, relying on interpretation and inference from incomplete data. In retrospect, his career seems to reflect a consistent pattern of challenging received wisdom for the sake of it, in the hope of greater glory. How else can we explain the title he gave his interpretation of an experiment involving four generations of sawflies: 'A New Principle of Evolution'? Although Heslop Harrison was not the only believer in an alternative theory to Darwinism, he certainly seems to have put a disproportionate amount of effort into proving a theory that, apart from extremely subtle variations of effect at the molecular level, is largely discredited today.

A rather sad sentence, buried on page 260 of the obituary, contradicts much of what Heslop Harrison's admirers had to say about him. In pursuing the Rum story, I found that even scientists who accepted that Heslop Harrison had done the dreadful deed would hasten to add that, of course, he was a very good researcher most of the time. But the obituary author's attempt to sum up Heslop Harrison's lasting contribution to science was the following: 'In their day these ideas excited much interest among biologists' (in the same way, I suspect, that the metal-bending activities of Uri Geller excited much interest among metallurgists), 'but today it is rare to find them quoted in standard textbooks, encyclopaedias and other works of reference published in English and in common use among students and teachers of all grades.'

And yet you can't escape the impression that it is those ideas he was passionate to be remembered for, rather than the slow, detailed, specialised collection of data that were meant to support the ideas and ultimately failed to.

My appetite was whetted by discovering doubts about this area of Heslop Harrison's work unrelated to botany, and I looked for similar examples. Another suspect claim of Heslop Harrison had been referred to in a letter in the Raven file: the story of the Large Blue butterfly mentioned by the Laird of Canna, the man with a fondness for flute quartets. When challenged by a journal editor, Heslop Harrison said that he had only meant to record the *possible* occurrence of *Maculinea arion* (Large Blue) on Rum. But a letter from Heslop Harrison in the Wilmott correspondence file, dated 26 April 1945, has the following postscript: 'In case you still cherish the famous(?) "Large Blue" yarn, let me tell you that captures have been made in the presence of well-known people not connected with us and that such specimens are placed in safe hands.'

Still living in Canna nearly thirty years later, Campbell continued to seethe over what he saw as an example of Heslop Harrison fakery. In 1973, fired to action by a footnote in an article about the Large Blue butterfly in Ireland, Campbell decided to track down the truth about the alleged sightings of this butterfly, which he believed had never been on Rum at all. The starting point for his quest was his memory of being shown a Large Blue that Heslop Harrison had given in 1945 to a David Parsons at Oxford. 'I expressed some scepticism,' Campbell wrote. 'The specimen looked old and dry and had no data label. The possibility that the Large Blue really occurred in the Hebrides seemed an extremely remote one to me.'

Campbell now set about trying to find the recipients of specimens, or witnesses to the alleged capture of the Large Blue, or *M. arion*. He told the story of his search in an article in *Entomologist's Record* in 1975.[54]

Three alleged specimens of Large Blue had been caught on Rum. First, Campbell found no trace of the specimen given to Parsons. Parsons thought he had given it to E.B. Ford, who denied possessing or even seeing it. A second alleged recipient, W. Campion, had died twenty years before, and no one knew where his collection was. The third specimen, Heslop Harrison had told Parsons, 'had been given to Oxford'. Campbell noted drily, 'But Oxford knows nothing of such a specimen.'

Then there was a story that a Miss Rhodes* had witnessed the capture of a Large Blue by the professor on Rum. Miss Rhodes was eventually traced; but because she was not a lepidopterist, she could not confirm its identity.

Campbell then tried to find where Heslop Harrison's butterfly collection had ended up – this was now six years after the

---

* The companion of Lady Bullough, who had shown Raven the relief map of the Isle of Rum.

professor's death. He eventually tracked it down and sent a friend, E.C. Pelham-Clinton, to inspect it. It was in such a poor state that it was 'worthless as substantiation of any of H.H.'s records'.

Campbell then wrote in his article:

It must be said that it would have been an extraordinary thing for a professional zoologist connected with such a well-established university to capture three specimens of such a rare butterfly as *M. arion* in a totally new and unexpected locality, and to have given two of them (one certainly unlabelled) away to amateur entomologists. An explanation of this might have been that Professor Heslop Harrison, having thought he saw *M. arion* flying on Rum in 1938, gave specimens from elsewhere to Parsons and Campion by which they might recognise the species if they visited the island. But this possibility is negatived by Parsons' recollection of his surprise at being given a supposed Rum specimen, and by the presence of Miss Rhodes at the time of the alleged captures.

The whole matter must be set against its circumstantial background . . . One has to consider John Raven's famous letter on 'Alien Plant Introductions on the Isle of Rhum' in *Nature* of 15 January 1949, which suggests that perhaps Professor Heslop Harrison's ardent and competitive personality may have laid him open to students' practical jokes.

John Morton, however, offered an explanation of why, if the Large Blue *had* existed on Rum at the time Heslop Harrison thought he saw it, it might now have disappeared. The explanation emerged during a visit he made to Rum with Heslop Harrison: 'We visited the dunes on which some years earlier he [Heslop Harrison] had collected the Large Blue butterfly. He said it appeared to have died out and explained this as due to the introduction of sheep on the island during the war. Apparently the owner had not permitted sheep on the island but maintained it as a wild hunting preserve for deer. Under these natural conditions the machair behind the dunes was not heavily grazed, and the ant with which the caterpillars of the Large Blue have a symbiotic relationship occurred in abundance along with its large anthills. The introduction of sheep, which was ordered by the government as a war measure, radically changed the habitat, with heavy grazing and trampling destroying most of the anthills, and the butterfly. Incidentally, similar problems have since led to the extinction of the Large Blue in its other British stations, in Cornwall and Gloucestershire.'

It has to be said that, for Morton, each suspicious instance of an unusual Heslop Harrison discovery had an alternative, innocuous explanation. 'The point should be made,' he told me, 'that in the many years I knew the Prof I never found any reason to doubt the many discoveries and new records of plants and insects that he had made. He was an exceptionally observant and diligent person in the field. This, coupled with his tremendous knowledge and experience, enabled him to find things where other people missed them. He was a *very good naturalist*.'

With the 1960s and 1970s having provided new published accounts of suspect data in Heslop Harrison's articles and papers, what would the 1980s bring? Well, what John Raven did for plants and John Lorne Campbell did for butterflies, Garth Foster was itching to do for water beetles.

The most persuasive thing about Garth Foster's account of Heslop Harrison's water beetle discoveries was that he had formed the view that these were fakes before he knew anything about Heslop Harrison's other dubious finds. There is always the possibility that if you know that someone has previously been suspected of fraud, you will begin to look more critically at his work and find 'evidence' that feeds your preconceptions. In Foster's case, he had begun quite early in his water beetle career to have independent

suspicions about certain of Heslop Harrison's reports, and the more
he looked into the matter, the more suspicious he became.

As with the plants, the dubious nature of Heslop Harrison's
water beetle discoveries surfaced in public in a very muted form,
couched in technical terms in an article titled 'Coleoptera in the
Inner Hebrides', as follows:

> I have omitted three of G. Heslop Harrison's (1938) early records
> from Raasay . . . Miller *et al* regard the record for this southern
> European species as 'probably doubtful'. Similarly *Noterus clavi-
> cornis* has not been recorded by subsequent collectors on this
> island . . . Miller *et al* listed a further nine of Heslop Harrison's
> species which have not been recorded since . . . since repeated
> searching by coleopterists has failed to find further specimens,
> this species is also omitted . . .[55]

It will not have escaped the eagle eye of the reader that the
Heslop Harrison referred to at the beginning of this extract is one
G. Heslop Harrison. George, another of the professor's children,
has not put in much of an appearance so far, but he has a starring
role in the water beetle story. Brian Silman, who succeeded
George at the University of Newcastle, described the department
to me as 'rather incestuous' when he joined it. What Silman
meant by this was that it tended to appoint only people who had
taken their first degrees at the university. But it could be seen as
incestuous in another sense. During quite a long period it seemed
to be rather full of Heslop Harrisons by birth or marriage. We
have already met Willie Clark, who was married to Heslop
Harrison's daughter, Helena, known as Dollie, who was also a
lecturer in the department. George Heslop Harrison became a
head of department at the university; his wife, Dorothy, also
known as Dollie, worked as a librarian in the Faculty of
Agriculture.

Garth Foster is an entomologist who first came across references to the Heslop Harrison family in the 1960s. Professor J.W. Heslop Harrison was one of a handful of field naturalists who took an interest in water beetles in the 1930s and 1940s, and Foster knew his name from the reports he published of significant discoveries in the Hebrides. When Foster moved to Newcastle University in 1967, he heard rumours that there was something suspicious about some of these records, but on the one occasion when he made reference to these rumours in the hearing of Dollie Heslop Harrison – the librarian – he was so put off by her angry reaction that he did not raise the matter again. But as he learned more about water beetles and talked to other coleopterists, it became clearer and clearer to him that some, at least, of the Heslop Harrison records could not be genuine.

As Foster outlined the story to me on the phone, it seemed that – rather late – I had come across a whole new area in which Heslop Harrison might have used the same dubious techniques of which he was suspected by botanists to persuade people that his biogeographical theories were correct. I decided that I'd better visit Foster and perhaps even see a water beetle or two myself.

Within moments of entering his stone town house in Ayr, I was being urged into a chair to look down a binocular microscope. I barely had time to take stock of Foster's crowded, busy study before I was confronted with a bright black-and-yellow beetle. It swiftly disappeared from view and another appeared. Before I had time to take in this one, blacker and larger, a third, smaller beetle slid under my gaze.

'You've got three quite distinct things there,' said Foster, lining up three of Heslop Harrison's most surprising water beetle discoveries in the Hebrides. 'That first one is ridiculous, in the sense that it's difficult to imagine how anybody else has failed to see it since. The second one is absurd in that it really is a true Canary Islands endemic. It's not found anywhere else in the world, so it doesn't make any sense that it should be found in mainland Europe, let alone in Scotland. And the third one is weird because it is just possible that it really is correct because it is so obscure. But if I was going to find it in Britain, I would not expect to find it on Rum. I'd find it on the mainland, first in mountains. And, again, nobody's seen it since.'

There, in a nutshell, were Garth Foster's reasons for believing that the most significant water beetle discoveries reported by Professor J.W. Heslop Harrison were hoaxes.

Foster is a coleopterist, a collector and classifier of water beetles. He works at a Scottish scientific institute where he runs a team that looks at the effects of pollution on biology and ecology, but the rest of his waking hours are spent with water beetles, his own records and specimens as well as beetles collected by others, which pour into his house in little packages by every post, requiring him to identify the contents and report back. Occasionally, amid the familiar beetles he knows well are specimens that require a little more thought and investigation. And, very occasionally, he has the excitement of coming across an insect that is new to a British vice-county, or to a British county, or even – if he's very lucky – new to Britain.

Foster is a short, animated man, with a passion for his subject, and the words spilled out of him as he helped me understand the background to the story of the Heslop Harrison water beetle discoveries. On a large table in his sitting room were a microscope and a computer, key tools in his task of being the recorder for all British water beetle discoveries.

'I am wet coleoptera,' he said. 'If I get a weevil, I send it to a man in Wales. If I get a leaf beetle, I might send it to a man in the British Museum. Ground beetles go to somebody else, and so on.'

Like many naturalists, Foster's interest developed as a boy, in his case with butterflies. 'I got fed up with butterflies very quickly and went on to carrion beetles,' he said, as if it was the most natural and obvious transition in the world. But even carrion beetles didn't hold his interest, and he then arrived at the insects that were to preoccupy him for almost the next forty years. 'People often ask me, "Why water beetles?" You might just as well say, "Why golf?" It's got me into every country in Europe and every ten-kilometre grid square in Britain and Ireland.'

He is quick to dissociate himself and entomologists in general from the much more numerous group of bird-watchers, often called twitchers, who are on the alert for any unusual bird sighting and rush off in droves to line up for a glimpse. 'We're not like twitchers,' he said. 'I like to say that entomologists talk to each other across the grave. If a rare beetle turns up, another coleopterist might turn up five years later and have a look, or it might actually skip a generation. People are born to this. It's very difficult to convert someone. It's not like a religion in that respect.'

My first surprise – after I had got my breath and seen the beetles presented to me under the microscope – was how small they were. Through the eyepieces it was impossible to know the scale. When Foster pulled out the square of plastic the beetle was pinned to, I could see that the insect was barely larger than a pinhead. 'We collect microdots,' he said. And next to the microdot I had been looking at was an even smaller speck. This turned out to be the genitalia of the water beetle, placed helpfully to one side as an aid to identification. There are thousands of different types of water beetle, and it can be difficult to tell apart some of the species from their external appearance alone. That's where having their genitals on show can help to differentiate one from another, assuming you have acquired the skill of removing the genitals from the microdot in the first place.

One of the species Foster showed me under his microscope was *Nebrioporus canariensis*, a species found in large numbers in the Canaries and never seen in a natural setting by anybody else within a thousand miles of the Hebrides. But it was reported by George

Heslop Harrison as a species new to Britain in a paper he wrote for the *Entomologist's Monthly* in 1936.[56] '*Canariensis* is abject nonsense,' Foster told me. 'To get two live *canariensis* into somebody's net in Barra in 1935 is inconceivable.'

George Heslop Harrison claimed to have collected sixteen specimens of different water beetles, on 10 October 1935, from a loch on Barra. On 22 October, these specimens were sent in a tube, preserved in alcohol or formalin, to Frank Balfour-Browne,* the Grand Old Man of insects, as Foster described him, and a friend, or at least an acquaintance, of Heslop Harrison's. In fact, it was the professor who sent the tube to Balfour-Browne, asking if he would identify the species gathered by his son. Balfour-Browne identified two of the beetles as a male and a female of the species *Nebrioporus canariensis*. That was as far as he could go since he relied on the word of Heslop Harrison that the insects had been found on Barra, but Balfour-Browne seemed confident enough to include reference to this beetle as an addition to the British fauna in Volume 1 of his *British Water Beetles*, published in 1940.

Foster believes that this discovery cannot be genuine, for several reasons. First, the extreme distance between the natural home of the beetle and the Hebrides, with no evidence of any other appearances between the Canaries and the Hebrides; second, that in the sixty years since this report, nobody has discovered *Nebrioporus canariensis* in the Hebrides, or anywhere else in Britain. (This is the second time the Canary Islands have cropped up in this story – the *Wahlenbergia* plant that seemed to corroborate the foreign origin of the *Carex bicolor* on Rum was a native of the Canaries.)

---

* One of the many odd twists in this story is that Balfour-Browne's wife is rumoured to have had an affair with Wilmott. Whether or not this is true, a close bond of friendship certainly formed between the two over the table tennis table.

There is a third reason for suspecting this discovery: the fact that in quick succession various members of the Heslop Harrison family came across other unusual water beetles that have never been corroborated either. In 1936, George Heslop Harrison reported discovering on the island of Raasay several specimens of *Aulonogyrus striatus*, a colourful and highly distinctive species of whirligig beetle never seen before – or since – in Britain. Again, Balfour-Browne was to accept this as a British beetle in his standard work, and again, Foster, considering the matter sixty years later, believes that it was a fake. Then, in November 1938, Balfour-Browne received another tube of sixteen beetles sent to him by the professor, this time, allegedly, collected by Heslop Harrison's other son, Jack, on Rum. This tube was found to contain five specimens of *Hydroporus foveolatus*, a small diving beetle found previously in the Pyrenees, Alps, and Balkans. To emphasise the firm hold the insect had on Rum, the professor sent Balfour-Browne three more specimens he claimed to have found there in 1939.

At the time I was visiting Garth Foster, I had not come across two very interesting letters from Balfour-Brown himself, to John Raven, stimulated by Raven's letter to *Nature* in 1949. These turned up in a cache of papers unearthed by Hugh Raven, John's son, early in 2016.

'Dear Sir,' the first begins, 'I have just seen a most interesting article by you in Nature . . .'

Balfour-Browne goes on in guarded terms to comment on Heslop-Harrison's activities with beetles. '. . . all sorts of extraordinary discoveries have been made by J.W.H.H. . . . There have been complaints from various entomologists that H.H. will not let them see things he has recorded. I am hoping that your cleverly shrouded implications may have a salutary effect . . .'

After Raven replied, Balfour-Browne wrote again, with rather less guarded comments:

H.H. has had plenty of time to protest against your insinuations if he intended to & since he has not done so, it seems to me to stand as evidence against him! In February, I heard a story in London that one of his students found him putting specimens into her (or his) collecting box and I am now rather inclined to follow it up and see if there's any truth in it . . . It is all very well for those who say "don't do anything about it" but how are we to study distribution if the material supplied is suspect? . . . I am particularly sorry that he should have descended to such a mean way of attracting attention as he has done more than almost anyone in taking his students into the field.

What became clear as Foster unfolded his story was the fact that a new element had been introduced: if the beetle records really had been faked, the whole business looked as if it had turned into a family affair.

Let's take George first. He was an entomologist and was about twenty-five when he went on the first expeditions with his father to the Hebrides. In the mid-thirties, his father had written to the Keeper of Entomology at the British Museum about a vacancy in the department:

My son would like to apply, but there are points that need clearing up. He took a B.Sc. (Pass) in Agriculture, specialising in Entomology. Instead of going for the extra year necessary for the Honours degree, he has spent 2 years researching on the Hemiptera.* He has entered for the Ph.D. and will take it at the end of this session. He is very good at the work and will certainly get the degree and do very well. How is he placed in regard to the post? Let me know and send me all further particulars. I am exceedingly anxious to get him an entomological post of the

---

* Bugs

nature of yours. He is a good technician, keen and good in the field and further is in love with systematics . . .

It is the sort of letter anyone who sees himself with an inside track will write in the hope of getting favourable treatment for his son, and it stimulated the sort of reply anyone would give who had no intention of providing that special treatment. 'From what you say,' the Keeper of Entomology wrote back, 'he would appear to be a strong candidate . . . The procedure briefly is that he has to write to the Archbishop for a nomination . . .'

George Heslop Harrison did not get a job at the British Museum but instead went to work in his father's department at Newcastle until 1939, when he was appointed Director of Plant Pathology and Entomology to the government of Iraq, a country that had been given its independence by Britain in 1930 but was still manipulated behind the scenes by the British government. In 1941, a pro-Nazi government staged a coup and many of the British residents left. George didn't, though, and as far as can be discovered, he was imprisoned as a British spy and tortured severely.

In a letter written in 1944, the professor says: '. . . although my son is medically "fit" (!!!!) <u>he is completely ruined for life</u>. The mutilations he received when he was tortured have left several severed ducts which surgical treatment cannot remedy.' In 2013, one of George's former students revealed that he had been 'electrically emasculated by the Nazi sympathisers in Iraq'. Ruined for life or not, George later became senior lecturer in the Department of Agricultural Zoology at the University of Newcastle and eventually its head. Accounts of people who worked under him at the time suggest that he was very difficult to work with – 'mad' was used by more than one person. Another of his former students described him as 'tall with thinning hair, with the stance of a praying mantis. He always wore a light blue Harris Tweed sports jacket with all the four buttons done up. He had yellowing, watery eyes and a chesty wheeze, which made him cough at regular intervals while talking.'

Of course, there is no reason at all to think that there was anything in George's character *before* his wartime treatment that would have led him to betray the science of entomology. Nor is there anything to suggest that his brother, Jack, named as discoverer of another of the suspect water beetles, would have lent himself to the faking of discoveries. In fact, Jack was a botanist, and we can expect water beetles to have been of less interest to him than plants.

But that suggests an explanation that, if correct, makes the professor's deeds even more heinous: he would have used his sons' good names to spread the load, so to speak, just as he was suspected of using Willie Clark as a cover for his own opinions. If the dubious specimens were reported by a range of people, it would make it less likely that accusations of fraud could be levelled specifically against the professor.

Both Foster and I found this an unpalatable explanation, but I asked him how the fraud might have been achieved, assuming that the sons were not in the know.

First of all, Foster explained, the specimens would have to have been dead when brought to the site. 'There's no way you could go to the Alps, bring back some live rare beetles, and put them in a net,' he told me. Using sleight of hand to pop a dead rare beetle in the net would have been unlikely to succeed – they would not have been noticed by George or Jack. The beetles were small enough as it was. With a dozen or more live beetles wriggling around, it's unlikely that a still black speck would have been noticed and included in the tube of specimens sent off to Balfour-Browne. 'My accusation,' said Foster, 'would be that Professor Heslop Harrison planted them on his son's sample. It would be a lot easier for the professor to wait and say, "I'll send those off to Balfour-Browne," then put them in afterwards.' And the fact is, in the case of the tubes containing the rarest specimens, although these were noted as having been gathered by George or Jack, they were sent to Balfour-Browne by the professor, along with a covering letter. But if the rumour was true that Balfour-Browne had heard, about Heslop-Harrison slipping specimens into a student's collecting box, it wasn't impossible that he slipped them into his son's too.

As Foster spoke of his conviction that Heslop Harrison had faked a number of water beetle discoveries, it became clear that, in his view, the three that he described in detail represented for him the worst examples of a practice that he believed was carried out on a wide scale. One aspect of the situation that confirmed his suspicions was that in these three cases, and with several other less surprising 'discoveries', no one had succeeded in finding further specimens since Heslop Harrison reported them. Barra, Rum, and Raasay, all sites of Heslop Harrison water beetle reports, have been thoroughly surveyed since by a variety of entomologists, with no trace of the rare beetles.

To get to the bottom of the problem, Foster himself carried out a survey in 1990 of a particular area on Rum originally surveyed in 1939 by Heslop Harrison, who then sent his collection of finds to Balfour-Browne for identification. Heslop Harrison reported twelve species of water beetle; Foster found fifteen. The two lists had only five species in common, meaning that Foster found ten species that either weren't there in the 1930s or were missed by Heslop Harrison, and Heslop Harrison reported seven species that were no longer there in 1990 or had never lived in that area. According to Foster, this exercise 'has undermined my confidence in the rest of the data, let alone the introductions to the British list'.

Once again, we have to ask, 'Why did he do it?' As we have seen, Foster believes that it was another salvo in Heslop Harrison's campaign to convince his fellow scientists that plants and insects had survived the Ice Age. In Heslop Harrison's view, the land-masses on the west of the British Isles had originally been connected, so that if plant and animal species had survived through the last ice age on the high, ice-free areas known as nunataks, they could have spread throughout the area across the 'land bridge' in the period before the ice melted, the seas rose, and landmasses like Ireland and the Hebrides were cut off.

Heslop Harrison's support of this theory as far as water beetles were concerned provided an example of a characteristic I was coming across regularly: his ability to make enemies. A stormy relationship developed between Balfour-Browne and Heslop Harrison over this very matter. Having acted perfectly amiably and professionally in helping Heslop Harrison identify the beetles he was discovering, by the 1950s Balfour-Browne had turned against him, as demonstrated by a series of public arguments about the land-bridge theory. The quarrel was exacerbated by the fact that Balfour-Browne also had a theory, which couldn't have been further from Heslop Harrison's. For Balfour-Browne, mobility was the answer. For him, insects that turned up in odd, unexpected places did so under their own steam or had been blown there by high winds.

From what we can infer about Heslop Harrison's likes and dislikes, Balfour-Browne was laden with qualities that one might expect would have irritated Heslop Harrison. He even had a genuinely double-barrelled name, derived from old Scottish families, unlike Heslop Harrison's fabricated one. 'He was toffee-nosed, rich, and stone deaf,' said Foster. 'He was also the only person to hold chairs in entomology at London, Oxford, and Cambridge.' Balfour-Browne's grandfather had made his money in railway law, and there was a family tradition of training in law. Balfour-Browne had trained as a lawyer but never practised. However, he retained an advocate's style of argument: 'When Balfour-Browne got stuck into you,' said Foster, 'you knew about it, and if you were a bit wobbly on your basic theory, then you suffered – and he would do it in public.'

Garth Foster lent me Balfour-Browne's copies of various issues of *Proceedings of the University of Durham Philosophical Society*. One, for July 1950, includes an article by Heslop Harrison titled 'A Dozen Years' Biogeographical Researches in the Inner and Outer Hebrides'. Pencilled lines down the sides of paragraphs; exclamation marks, sometimes in triplicate; and marginal comments litter the article. 'Pure guesswork', Balfour-Browne has written by one paragraph, 'nonsense' by another, and a detailed note in tiny handwriting points out that you can't always rely on local traditions to bolster your theories.

An issue that almost brought the two to blows concerned the mobility of insects. Heslop Harrison had teamed up with another researcher, Dorothy Jackson, who had demonstrated through dissection that many of the wandering beetles didn't have flight muscles, supporting Heslop Harrison's view that they could not have flown from island to island but must have crossed by land.

In a book rather unimaginatively titled *Water Beetles and Other Things*,[57] Balfour-Browne tells of a meeting at the Linnean Society where his theories and Heslop Harrison's, as purveyed by Miss Jackson, clashed:

Professor Heslop Harrison's opening remarks reminded me of the legal cliché 'If you have no case, abuse the opponent's counsel' as, having asserted that 'every pertinent fact, botanical and zoological alike, led him to the same conclusion as that put forward by Miss Jackson', he remarked upon my 'complete lack of knowledge of the habits of the Belted Beauty and of its distribution'. I said that the male moth might have carried the female, during coitus, across to the islands. His objections were that the journey would have been 'against the prevailing winds and at colossal speed'. He forgets the prevailing winds when he writes about the butterflies which migrate to the islands and he appears to assume that the paired moths would fly across in calm weather. If they crossed in the way I suggested it was because they were caught in a gentle updraught of air which gradually increased in strength with elevation and a favourable wind carried them across. His remark as to the length of time the pair remained connected does not matter since, once the two moths were uplifted, the wind would carry them across whether or not they were as one. Thus even the colossal speed he suggested is unnecessary. I have never handled the Brindled Beauty moth but when Harrison was about seven years old I had a large family of Vapourers and my recollection of their coitus is that it was somewhat drawn out.

In a summarising paragraph, Balfour-Browne gritted his teeth and actually made a favourable comment on Heslop Harrison before letting loose a final barb:

Harrison has done really excellent work on the fauna and flora of these western islands and his mistake has been to jump to conclusions on insufficient evidence. He admits some mistakes but the fact that he has made them makes it possible that there are other mistakes that have not come to light.

Foster is clearly a Balfour-Browne fan. The yellow T-shirt he sported when he met me at the airport had BALFOUR-BROWNE CLUB emblazoned across it, and at his home he proudly showed me a garage bill of Balfour-Browne's from 1937. 'It's got three stamps on it, with the heads of three kings,' he said, pointing to George V, Edward VIII, and George VI.

There was one final element in the water beetle story that Foster suggested – somewhat mischievously – might have been a factor: 'We're almost getting to a stage where we could say, "This is Heslop Harrison sending up a smoke screen for Balfour-Browne, since he saw most of this material." I think he just said, "What shall we send that daft bugger next? How can we wind him up?"'

As we finished our discussions Foster said, harking back to an earlier remark, 'We're strange people, you know. We're not golfers.'

If we add Garth Foster's beliefs about Heslop Harrison's water beetle work to John Lorne Campbell's Large Blue investigation and John Raven's Rum research, we have three resounding blows to Heslop Harrison's reputation as a scientist. Along with the cloudier matter of the moths and the sawflies, they leave little doubt in my mind that Professor J.W. Heslop Harrison was a man whose scientific research had reached the stage where his results and data were just no longer to be trusted. How long, then, did it take the professional world to realise that, whatever the rights and wrongs of accusing him publicly of fraud, his data must not be allowed to stand?

12

The Matter Subsides

At the time of the Rum affair, Raven had just been made a fellow of King's. After a year he was made Lay Dean, a college post that superficially had to do with minor disciplinary matters. King's was a college that, like Raven, seemed rather embarrassed at having to enforce rules on staying out after midnight, wearing gowns, and walking on hallowed areas of grass. But the Lay Dean was also the don who was expected to make a particular effort to get to know the students, and this became more important as King's broadened its intake to include more state-school boys (of which I was one), partly at Raven's instigation. His interest in botany continued, and among the papers Hugh Raven discovered early in 2016 was evidence that John Raven was still concerned about the story, perhaps irked by the fact that no public action had been taken and that many of Heslop-Harrison's 'discoveries' were still included in the reference books.

Enclosed in a letter from a Cambridge solicitor to Raven in November 1951 is a Counsel's opinion 'In the Matter of Mr. John Raven' written by a barrister, John Davidson. This appears to be a comment on a revised draft of a reply to some comments by Heslop-Harrison which Raven had submitted to the solicitor, and although the revised version is 'much more objective than the original' Davidson still has some nits to pick.

Page 2. Delete the word 'now'.
Page 4. For 'proved' write 'shown'
Page 7. For 'demolish' write 'discount'.

And so on and so on.

The paper in question may have been written for a journal called *New Phytologist*, edited by Arthur Clapham, who was originally meant to be a member of the Rum expedition, along with Maybud Campbell. Two weeks after Raven received the libel report, Clapham wrote to Raven about 'your paper' and in spite of the fact that he knew as much about Heslop-Harrison's misdeeds as Raven himself, he said in his letter that he thought Raven's article was 'unnecessarily polemical', and – in a move which would have been unlikely to endear him to Raven – he submitted a version, presumably less polemical, which he had written himself. There is no evidence that Raven accepted this redraft, and in a second letter Clapham says 'I am worried about our difference of opinion. I had hoped to see you at the Cambridge Press sherry party on Thursday so that we might have a talk together . . .'

One can imagine how Raven felt, possessed of glaring evidence of Heslop-Harrison's frauds, as people whittled away the case he was making until it was in danger of disappearing up its own argument. In fact, no imagination is needed, because enclosed with the lawyer's report is a handwritten and heartfelt reaction to the lawyers objections, finishing up with: 'Have sweated blood over trying to make this as objective and innocuous as possible. Time unfortunately, is an important factor: the longer my reply takes to appear, the less chance it has of achieving complete success, i.e. the deletion of many plants from the official list.'

Six months after this final flurry of correspondence, on 7 May 1952, Raven received out of the blue an extraordinary letter from Heslop Harrison. Like two earlier letters, it was written following a spell of personal sadness or ill health and seems, in this case, to have been triggered by it. Its dramatic heading is 'Solely for your eye and conscience':

Dear Sir,

My wife has just died after a long and painful illness of over four years. I feel that I must inform you of the pain and unnecessary suffering your lust for notoriety caused her.

In 1948 I took her to the Isle of Rhum for her health's sake, and, most unfortunately, she received the first impact and most serious consequences of the trouble that arose owing to your creating the impression that I was responsible for your appearance on the island and for your bringing a wholly unauthorised person who, by the way, you informed me was no botanist (but was quoted in an attempt to bolster up your empty statements) with you. Moreover, your actions caused us to lose our rooms at the Castle. As a consequence she came home much worse than when she left.

Again, most unfortunately, she was the first to become acquainted with your cowardly letter in *Nature* with the sly smear implied in the unwarranted introduction of my name. Fortunately, her informant did tell her that the attempted smearing was baseless and due to your pandering to your personal vanity. At her wish, her knowledge was kept from me; nevertheless, she fretted about it and it greatly increase [*sic*] her sufferings. I only saw the dastardly production only [*sic*] in 1951 when a student told me about it.

I want to tell you plainly that what you wrote was a complete fabrication cemented together by a few facts, often distorted and their source never acknowledged, obtained from me. Further, I must inform you that several competent botanists in my company thoroughly searched the Black Corrie for *Sagina apetala* and *Juncus capitatus* in 1949 and 1951 and were unable to find them. In my opinion they never occurred there.

Again, I cannot understand the mentality of anyone who

accepted the kindness and hospitality of my wife and myself, and even had the impudence to shake hands with us when he left when he knew the ghastly brew he was concocting with the intent of adding to his own importance.

I want no answer to this letter and any attempted reply will be burnt unread for I have had abundant examples of the utter worthlessness of your work. My sole wish is that my letter will prevent your desire for aggrandisement causing pain and distress to other equally innocent people in the future.

One additional point I should make is that I have further evidence of your activities in other directions with the same general intentions of smear and additions to your self-importance.

Yours sincerely,

J.W. Heslop Harrison

Two letters are in my possession, one proving that you planned to be on the island when I was there and the other proving that you were 'after' the two sedges.

Tom Creighton said that John Raven found this letter 'rather painful'. Whatever Raven thought of Heslop Harrison's professional misdeeds, the letter clearly shows a man in great distress, thrashing around for any stick with which to beat him. The story of the Heslop Harrisons losing their rooms at the castle had already been told in the previous correspondence, and it is difficult to see anything in Raven's account of the events on Rum that would have led to such an expulsion, particularly if Mrs Heslop Harrison was as ill as the professor implied. Similarly, the dark hints of secret knowledge of Raven's other activities are unsupported by anything else I have been able to find. I can't even imagine what Heslop Harrison is referring to.

In 1954, Raven married Faith Hugh Smith and the couple made their home at Docwra's Manor in Shepreth near Cambridge. With a garden that they started to fill with plants, and with the estate of Ardtornish in Scotland that Faith brought to the marriage, Raven's botanical interests had new fields to satisfy them. His classics teaching and research continued, and he wrote several introductory books on classical philosophy. When no university appointment was forthcoming – unlike Heslop Harrison, he never became a professor – Raven turned his attention increasingly to his beloved botany.

By the time the elder Heslop Harrison died, in 1967, there had been only the most muted of public criticisms. He seems to have had a comfortable retirement, still taking an interest in his favourite aspects of natural history. According to his Royal Society obituarist, 'His last few years, when his health was failing, were made very happy by kind friends and relatives who took him by motor-car to his favourite places where he had found his moths, his willows and his roses.'

There was no rush to judgment after Heslop Harrison's death, and the nearest his obituarist came to referring to any unpleasantness was a footnote that said, 'Information on a number of Harrison's controversial records is embodied in confidential papers deposited in the Department of Botany, British Museum (Natural History).' Of course, I looked for these papers but at the time no one in the museum was able to find them.

In 1980, the year John Raven died, it was decided that an indispensable work – indispensable for sedge fans anyway – *Sedges of the British Isles*, should be revised. Originally written by Clive Jermy and Tom Tutin in 1968, the book was in enough demand to go through several editions up to 1977. For the revision, Jermy sought the help of some colleagues and sedge experts, including Dick David.

David was an old friend of Raven's and coedited the privately printed book *John Raven by His Friends*. For many years he worked at, and eventually helped run, Cambridge University Press. He was also a keen botanist, so he took a particular interest in Raven's Rum exploits.

Faith Raven recorded an interview with Dick David in the 1980s, in which he summarised his memories of discussions with

Raven about Heslop Harrison and added several new – and uncheckable – facts to the story:

> He obviously was a very masterful person and liked to impress his students. And there are stories [from] the expeditions that the professor would lag a little behind the rest of the party and would suddenly shout, 'I have found a plant new to Britain.' The students would crowd back and be rewarded by being shown a plant which, we suspect, the professor had at that moment inserted into the ground. The reason why it has taken so long to get to the bottom of this affair is perhaps a double one. In the first place, during the war, the Hebrides had of course been out-of-bounds to British botanists. Heslop Harrison therefore had a clear field, and he was the only man who could be said, at that time, to know much about the territory. Secondly, it is only recently that we have become familiar with the ecology of the high alpine species, and therefore it is only now that we can say that it is an absurdity to suppose that plants whose particular habitat is above three thousand feet in snow patches should appear at fifteen hundred feet or lower in the Hebrides where snow practically never lasts more than twenty-four hours.

When revising *Sedges of the British Isles*, the task fell to David of deciding how to refer to Heslop Harrison's 'discoveries,' the subject of increasing comment among botanists, or at least among those who concerned themselves with the flora of the Hebrides. Should *Carex bicolor* be left in the reference books on the basis of Heslop Harrison's official, and usually unchallenged, reports in the journals? If not, how could the issue be dealt with tastefully, and who should be consulted? As a close friend of Raven's, David had read his full report, but since it wasn't a published document, he couldn't very well use or quote from it, although there was always the *Nature* letter. David decided to see Heslop Harrison's son, also, confusingly, Professor John Heslop Harrison, and his wife and ask them what he should do. (He was the 'Jack' referred to in the story of Heslop Harrison's water beetle expedition.)

Jack Heslop Harrison had become a distinguished botanist in his own right and was Director of Kew Gardens at the time. His wife turned out to be the lady who played a cameo role in Raven's report as 'Iolande' (in fact, 'Yolande'), whose vasculum was being used to store sandwiches. During 1997, I made several attempts to contact Professor and Mrs Heslop Harrison myself, writing to them both by recorded delivery and receiving no reply. I then decided to make contact with Heslop Harrison's grandson, also a botanist and also a professor.

Pat Heslop Harrison works at the University of Norwich, at the John Innes Centre, and is the botanist who Dr 'O'Connor' feared would veto grant proposals if he spoke to me on the record about Heslop Harrison the elder. I wrote to Professor Pat in fairly general terms, saying that I was writing about his grandfather and asking to visit him, and received an amiable reply saying that he wasn't sure that he could help very much but if I wanted to see him that would be fine.

On a sunny, hazy day in May 1998, I drove to Norwich for my first encounter with any member of the family of the man who had been at the centre of my researches for the past year, and whose tainted reputation would be widely publicised (anyway, I hoped widely) by what I wrote. Pat Heslop Harrison greeted me warmly in the reception area and led me to his small office in the low, two-storey building where he worked. I sat in front of him and started to tell him how I had come across the Raven report into aspects of his grandfather's researches and that this had led me to explore further the allegations against him.

Pat Heslop Harrison looked slightly puzzled and apologetic. 'I'm sorry,' he said. 'I'm not quite sure what you're talking about.'

Had he never heard of a document written by John Raven in 1948? I asked.

'No,' he said.

He'd heard of allegations of faking results and 'planting' plants levelled against his grandfather, surely?

'No, never,' he said.

I could hardly believe what I was hearing, yet Pat Heslop Harrison seemed so open and honest – and, indeed, his denial was so unexpected – that it had to be true.

Professor J.W. Heslop Harrison had died when Pat was six, and clearly such matters would not have been aired in front of a child. What I found difficult to believe was that as the grandson became a botanist – 'I was interested in biology but didn't like the dissection,' he told me – he hadn't heard any of the stories that must still have been circulating among older botanists. And there was another source of such stories I would have expected him to hear from.

'Didn't your father ever tell you anything about it?' I asked.

'No, never,' he said. 'By the way, my father died two weeks ago.'

This was obviously to be my day of surprises. Here was I, embedded in my research, and I hadn't even noticed that a key individual in my story had just died.

The rest of the morning was spent filling Pat in on the details of the allegations against his grandfather, and eating a pleasant lunch in the John Innes cafeteria. Pat Heslop Harrison's main interest in the story was as a scientist whose own research leads to discoveries and published papers and, he hoped, new theories, and he therefore felt it was crucial that the highest standards be upheld. (Part of his research is in the genetic improvement of oil palms, to increase their yield of palm oil, used in chocolate to make it snap in pieces.) He gave no hint – as, indeed, I had always felt there should not be – that he felt the members of the elder Heslop Harrison's family should see themselves as sharing somehow in his guilt.

But apparently, from what I can gather, Jack Heslop Harrison had never been able to talk openly about the affair. After Dick David went to consult him about what to say in the new edition of *Sedges of the British Isles* concerning his father's claims, this is what he wrote, in a final section of the book headed 'Dubious Records', after listing *C. glacialis, C. bicolor*, and *C. capitata*:

> The first two were recorded for Rhum, v-c. 104 (Heslop-Harrison, 1941 & 1946) and the last mentioned species for S Uist in the Outer Hebrides, v-c. 110 (Heslop-Harrison, 1946) as a single tuft. All have since disappeared from these localities and we consider them to have been planted.

That one word – *planted* – such a botanical word for an unbotanical act, was the nearest anybody in the field came in public to accusing Heslop Harrison of fraud.

Mary Briggs, President of the Botanical Society of the British Isles, believes that the entry in *Sedges of the British Isles* was written with the agreement of the Heslop Harrison family. But she makes the interesting suggestion that even David might have needed a tiny bit of reassurance about the verdict. She takes groups of botanists to interesting parts of the world, and when David heard that she was planning a botanical trip to Switzerland, he asked whether he could go with her to see the natural habitat of *Carex bicolor*. Briggs consulted Swiss botanists and discovered a site that she suggested they visit. When they got there, she said, David was relieved to find that the habitat was very different from the site where Heslop Harrison claimed to have discovered the plant on Rum.

It seems that during the 1950s and 1960s, as the father's scientific work declined and the son's rose, botanists who knew of the story watched carefully to see if there was any indication of what the son knew or was willing to reveal. One year Jack Heslop Harrison wrote a paper about geese as carriers of seeds to Britain from Greenland. Was this some kind of signal, perhaps? Then, sometime in the late 1950s, the British botanical world became very excited at a BSBI conference. As Max Walters described it, Jack gave a paper titled 'The Significance of Botanical Records in the Outer

Hebrides'. 'Most people believed that Jack went to his father and said, "This is your last chance",' related Walters, 'and he delivered a paper that significantly ignored all of his father's records.'

John Morton also knew Jack Heslop Harrison and doesn't find it surprising that he was reluctant to become embroiled in discussing his father's suspect activities: 'I found Jack to be a very aloof, superior, and unapproachable sort of person. The professor's other son, George, an entomologist, I never got to know. As to why his sons did not rally to the Prof's support in public, I think the answer's obvious. They had to start their careers and make their own reputations at the time that the accusations against their father were rampant. To have written or spoken in support of their father would have done no good to him – people would dismiss such support as the natural thing to do for a son. It was important if they were to make a successful career that they distance themselves from their father and, of course, from the accusations. This they did, and very wisely in my opinion.' Jack's account of his relationship with his father, posted on a Heslop-Harrison website, is consistent with all we have learnt of the Prof:

> [From the age of 7] my father increasingly dominated my life in one way or another.
>
> His own boyhood had been very hard and rough, and although he had parental encouragement, especially from his mother, the financial situation of the family was such that any advance he made through the educational system of the time had to depend primarily on his own efforts in gaining bursaries and scholarships.
>
> He had in consequence developed a rigid mental discipline so far as work was concerned, and this he sought to impose on his children, for whom both he and my mother had high academic ambitions.
>
> As the last of the sequence, I felt pretty continuous pressure to achieve, and anything less than first position in my class in any

subject (especially in science) was regarded as tantamount to failure.

Regardless of whether anyone was willing to criticise Heslop Harrison publicly, there was clearly a duty to reflect the doubts about his discoveries that were rife in the botanical community, and so as new reference books came out it was interesting to see how the authors had dealt with this delicate question. *Flora of the Outer Hebrides*, for example, was edited by Richard Pankhurst and J.M. Mullin, and published in 1991 by the Natural History Museum.

I spoke to Pankhurst once, in the early days of my researches, and he indicated that he had little to add to what was in the book, although he did say that Heslop Harrison documented his discoveries in a very cavalier fashion and often didn't supply specimens but when he did 'some must have been falsified'. He didn't seem very concerned about the whole business: 'You can look at it statistically,' he said. 'All but a few of his plants were genuine discoveries.'

When I rang him several months later with some botanical queries based on my reading of *Flora of the Outer Hebrides*, the message came back that he did not wish to talk to me at all. Like O'Connor, Pankhurst presumably found the whole area too sensitive to deal with, in his case even after the death of Jack Heslop Harrison.

But if you know what you're looking for, *Flora of the Outer Hebrides* still makes things pretty clear. Here are some of the considered views of Pankhurst and his collaborators about the records of discoveries by one of Britain's leading botanists of the first half of the twentieth century:

> Prof. Heslop Harrison's preliminary Flora is comprehensive, but is also an annotated checklist. Its usefulness is reduced by the fact that a few of the records are known to be inaccurate, which unfortunately casts doubt on the many probably genuine but unconfirmed records there published.[58]

Unfortunately, many of the specimens presumably collected by Prof. J.W. Heslop Harrison have never been located. Some of his duplicates are at Kew and at Cambridge, but the whereabouts of the bulk of his collection remains a mystery. Part of the collection of his colleague, W.A. Clark, who took part in many of the Newcastle expeditions to the Outer Hebrides, was donated to the Museum, but unfortunately this proved to be mostly too decayed to be of any value.[59]

The report of the nut of *Trapa natans*, the water chestnut . . . (Heslop Harrison and Blackburn, 1946) would indicate an enormous northward range-extension of *Trapa* and suggest[s] a major climatic warming. The nut was not, however, found *in situ*, but washed up on the loch shore . . . In the absence of further work, this record remains enigmatic.[60]

There was, sadly, a degree of controversy with regard to some records, making the work less valuable than it might otherwise have been. However, many of the allegedly doubtful species have more recently been re-recorded, and the work of Heslop Harrison and his team remains a major contribution which cannot be ignored by any serious student of the Outer Hebridean flora.[61]

Following these remarks in introductory essays, there are then very specific cautions about the confirmability of Heslop Harrison reports scattered through the accounts of individual species. For example:

S. marina
Herniaria ciliolata Melderis
This is a rare plant of Cornwall and the Channel Islands. It was recorded by J.W.H. Harrison from South Uist, golf course at Kildonan . . . This species is very unlikely from this locality, and has not been seen there again.

Illecebrum verticillatum

... recorded from Barra by J.W.H. Harrison ... Requires confirmation; the habitat in Barra is now quite unsuitable ...[62]

G. pusillum L.

Only recorded by Harrison ... from South Uist ... and Lewis, Stornoway, 1956. The records may be errors for the last.[63]

So according to Pankhurst, Heslop Harrison may have been making 'errors' at least up to 1956, eight years after the Rum events.

Heslop Harrison was a particular expert on roses. But even in this category, Pankhurst and his colleagues are less than complimentary about the records produced by the Newcastle team:

'... Harrison and Bolton published what, at the time, was a definitive account of Roses in the Hebrides, but we cannot now approach the subject with such confidence ... When the rose specimens in the herbarium of W.A. Clark were examined they were found to be so badly decayed that none of them could be named.'

And so the roll call of Heslop Harrison reports goes on: '*Crassula* ... may be an introduction ... *E. pratensa* ... no specimens have been seen ... *Polystichium* ... the literature records were rejected ... *D. carthusiana* ... J.W.H. Harrison recorded it but there are no specimens ...'

The official verdict of this HM Stationery Office publication is the nearest the profession has got over the years to any public revelation of the inadequacies of a leading British academic over a period of thirty years in the middle of the last century.

13

'Broken, Lost or Never Collected'

As I approached the end of my researches, I felt that I had gone as far as I could in pinning down the facts of the Heslop Harrison story. Not that I *had* pinned them down, and not that there weren't more people I might have seen or contacted who had opinions, views, and hypotheses. But I had reached the point where each new piece of information I gathered was adding less and less to the picture of the complex man who was Professor John Heslop Harrison, and what little was added was consistent with the overall picture I had already formed. But to my surprise, in spite of jumping through all the right hoops to ensure that I had found all the archival material that existed in the Natural History Museum archives, two new 'caches' of documents came to light, one shortly before I completed the first edition of this book.

In all my researches I had come across hardly anything that suggested that Heslop Harrison was aware of what people were saying in private about his research until quite late in the story, when he mentioned to Wilmott the Large Blue 'yarn'.

Then, as I was gathering a few loose ends of research, I rang John Thackray, the archivist at the Natural History Museum, and he said, 'Oh, I came across a file the other day that I thought you'd want to see. It's marked "Harrison" rather than "Heslop Harrison", which is why we missed it when we were looking before.' I asked him what was in the file, but he said he was only looking at the catalogue reference, which said that it ran from 1935 to 1978.

I remembered the footnote in the Royal Society obituary of Heslop Harrison and wondered if, by any chance, this could be the 'information on a number of Heslop Harrison's controversial records . . . embodied in confidential papers in the Department of Botany'. In fact, the file Thackray mentioned was in the entomology archives rather than the Department of Botany and contained correspondence between Heslop Harrison and Norman Riley, Keeper of Entomology. I made an appointment to see it.[64]

People often used words like *domineering, dictatorial*, and *authoritarian* about Heslop Harrison, never *vulnerable, cornered*, or *lonely*. But the letters I found in the file were in some sense a pouring out of the heart to someone whose friendship and support Heslop Harrison needed and valued. Along with Heslop Harrison's expected anger, petulance, and bluster was a more vulnerable tone, expressing the surprise of a man who, because of his domineering nature, had escaped criticism for much of his working life and was now hearing it for the first time, and it wasn't pleasant.

And embedded in the correspondence was some analysis and some advice that, if heeded, might have changed the course and the nature of his research for the better.

The correspondence took place in 1939, long before the Ravens, father and son, had embarked on their quest. It arose out of the claim made by Heslop Harrison that he had seen or captured a Large Blue butterfly on Rum, the topic thoroughly explored by John Lorne Campbell in his article in 1975. It appears that Heslop Harrison had written a paper on Hebridean moths and butterflies that he had sent to Norman Riley, who had an editorial role on the journal *Entomologist*. It's typical of the ambiguity of the Large Blue affair, as Campbell learned, that in the Riley/Heslop Harrison correspondence we find Heslop Harrison withdrawing his earlier claim to have seen the butterfly 'because we might go back to Rhum'. The implication seems to be that if it didn't look as if he was going to return to Rum, it was worth mentioning in print that he had seen the butterfly, whereas if he was going to get back to the island, he would be wiser to hold his fire until he had captured a specimen.

In any case, in writing to Riley about this, Heslop Harrison had wanted the matter to be kept quiet. He was therefore very upset when he found that word of his original claim had got out, and he complained bitterly to Riley in a letter written in April 1939, in which he complained that someone must have been shown, or told about, the contents of his original paper, claiming to have seen the Large Blue.

I couldn't find Riley's reply to this letter but, whatever it said – and Heslop Harrison in his reply sent on 9 May thanked Riley for his 'friendly letter' – it unleashed a torrent of anguish from Heslop Harrison that extended beyond the issue of the Large Blue to various other unspecified accusations that had clearly been making the rounds for some time:

> I have been supplied by friends with the names of four of the persons concerned in making the statements of which I complained. As these statements are absolute lies I am going to take legal action on the first occasion they are repeated and I receive information.
>
> I assert without fear of contradiction that if any specialist states that he could not examine our material because it has been 'lost', 'destroyed' or 'never collected' then that person is telling a malicious lie. In this connection you say, 'To uncharitable souls such facts lend themselves to a very ready misinterpretation not to your credit.' As I say, the answer to that is that they are not facts they are wicked lies.

Heslop Harrison goes on to describe the care he has taken in the past to send material in good condition to the major centres of botany or entomology, and to complain that some of these were never acknowledged by Riley's own departments and that therefore he would send his specimens in future to other centres.

Finally, he writes:

I really cannot close without protesting finally against the rottenness of the individuals who would set out so maliciously to injure someone who has never injured them in any way and whose sole interests have been to advance the interests of science and who has never refused help to a single individual who has needed it or asked for it. The guilty persons must have minds of an order which an honest person cannot understand! The word 'doubter', which you use, about such people very much more than flatters them. Let me assure you that this letter is intended to be a friendly one and that sign [*sic*] myself in all sincerity.

Yours very truly

J.W. Heslop Harrison

One of Heslop Harrison's characteristics when he was upset was to use the pursuit of one argument as an opportunity to bring up other grievances, however unconnected. We have seen this happen in his correspondence with Raven, in which he returned time and again to the minor issue of Raven's discourtesy in the midst of defending himself against the major charge of fraud. Similarly with Riley, Heslop Harrison complains about the lack of acknowledgement of previous letters and specimens in the midst of his unfolding anxieties about the charge summed up in the phrase 'lost, broken or never collected'. Like 'Is such a thing done?' this phrase becomes a litany as the correspondence continues. Unfortunately I can find no letter *to* Heslop Harrison that contains this accusation, and it is not clear how or where all these rumours were circulating. But circulating they clearly were.

Riley obviously thought long and carefully before replying and giving the comments for which Heslop Harrison had asked him. One reason his reply needed careful handling was that Riley knew he had to reveal that he was partly to blame for Heslop Harrison's first complaint, about the leaking of his Large Blue report. And he cannot have been unaware of the effect this revelation might have, since it involved one of Heslop Harrison's bitter enemies.

On May 13, 1939, Norman Riley wrote to Heslop Harrison:

My dear Harrison

First . . . I have a confession to make. It was at lunch one day that Edelsten and I discussed your note about *arion*, and Wilmott was with us. He has been a close friend of mine for many years, and yesterday I asked him whether I had ever mentioned your report to him. He reminded me of the above circumstances and also pointed out that he believed it was he himself who suggested the report should be held over, for the reasons given in my letter of October 20th, 1938, to you. I now regret that I did not publish your note as it stood, as there could then have been no argument as to what you did or did not do or say. Even now if you still have the MSS. it might be a good thing to publish it. Wilmott assures me that he has not discussed the *arion* report with any entomologist; so far as he can recollect he has only mentioned it to Clark (who first saw it?) and to Miss Campbell. I know we agreed not to broadcast the report.

My recollection of your note is that you conveyed the impression that you were pretty sure that the insect was *arion*, but that you could not record it definitely because you had not caught any. How confident you were originally about it is shown by the following postscript from your letter of July 28th, 1938, to Edelsten. 'The most extraordinary thing here is *M. arion*. We have seen two but did not catch them.' Now about the 'broken, lost or never collected fable'. The thing isn't so simple as you think; I wish it were.

Riley then quotes from some of Heslop Harrison's own letters which revealed a certain degree of carelessness and sloppiness in the handling and sending of specimens.

> Now [Riley continues], all these inquiries concerned material of a critical nature. I have, myself, no doubt that the material concerned was not available simply because it had not interested you to see that it was carefully preserved. To most of us here it is almost inconceivable that anybody should fail to preserve <u>with special care</u> material of such interest and importance. Now, suppose for a moment that you had discovered that one of your most interesting records had been based upon an error of determination, and that you were thoroughly unscrupulous and determined not to admit the mistake. What would be the obvious thing to do? Why, destroy the material, of course. Don't you see how you have laid yourself open to criticism through allowing your material to pass out of your control; how easy it is for anybody who doesn't know you to draw a false conclusion, and, indeed, to do so honestly, without malice at all? Take the three extracts I gave you above: in the first case the species concerned proved to be (as you said) a common one though till then overlooked, but in the second and third cases, so far as I know there has been no further development.

It is difficult to imagine how Heslop Harrison would have reacted to this last paragraph, the heart of the letter and a skilful and diplomatic way of pointing out how Heslop Harrison's actions were consistent with those of someone who is faking his discoveries.

To deal with Heslop Harrison's accusations of letters and specimens being unacknowledged, Riley supplies a long list of refutations of individual complaints. He had obviously spent the last few days getting his colleagues to dig into their files and come up with copies of their own correspondence with Heslop Harrison. Riley concludes with another gently worded accusation, this time directed at Heslop Harrison's memory and at his record-keeping activities.

> To sum up. All the quotations I have given you above are from your own letters and are in my department. To me it is quite clear what is happening. You have no copies (presumably) of these letters to refer to, and your memory is playing you tricks; it is so easy to let 'what you intended to say' become 'what you said' in the course of time; and to forget the <u>actual wording</u> of a letter which may have meant one thing to you and something different to the recipient.
>
> If, finally, I may offer you a word of friendly advice – forget it all. You will be getting an 'encirclement complex' if you are not careful, whereas all we want of you here, and offer you, is frank co-operation.
>
> Yours very sincerely
> N.D. Riley

Even John Morton, Heslop Harrison's principal defender today, expressed to me the same sort of criticisms as those written by Riley: 'A major shortcoming as a scientist in my view was his failure to keep good and properly documented specimens of the plants and insects that he found and the ones that he worked on. He just did not have the patience to spend the necessary time to do this essential part of research. As a result he left himself open to the criticisms of those who questioned his discoveries. The adequate documentation of those discoveries in many cases just isn't there, and so the criticisms easily stick and will probably go on forever. It's sad. He was a great man, and so much of what he achieved is in question.'

It would have been surprising if Heslop Harrison had read Riley's letter in the spirit that was intended, and even more surprising if he had actually taken Riley's advice to 'forget it all'. So, of course, he didn't. Chapter and verse must be countered by more chapter and verse. And far from forgetting it all, in his final letter in this sequence, Heslop Harrison allowed his anger at his accusers to reignite. This letter was written on 25 May 1939, en route to the Isle of Coll:

> My dear Riley,
>
> I picked up your letter as I left Newcastle today and I must say that its contents surprise me. They make the whole episode still more disgraceful to its concoctors. Instead of being the simple family party as portrayed by you, I have had the wretched story from several sources.
>
> You state that the 'broken, lost or never collected' fable isn't as simple as I think; you never said a truer thing!! The stringing together of things wholly disconnected in an endeavour to link together something to justify the statement won't do . . .

Heslop Harrison then takes each of Riley's references to his colleagues' accounts of their dealings with him, and picks them apart, displaying the same affronted self-justification that he was later to use with Raven and others, continuing with:

> . . . as for your concluding paragraph on this shocking concoction let me single out one sentence. 'To most of us here it is almost inconceivable that anyone should fail to preserve <u>with special care</u> (underlining yours) material of such interest and importance.' Who says that the material was not preserved with due care? No one except someone who wishes to uphold a concoction now kept going for very different reasons!
> . . . No, that particular cock will not fight! It is not only minus spurs but hasn't even the consistency of a disembodied spirit!

When I first started talking to botanists about Heslop Harrison, one commented on the oddity – even tastelessness – of Heslop

Harrison's sporting a Hitler moustache during the Second World War. I never dreamed that he might be indirectly reflecting remarks that had been made sixty years ago by Heslop Harrison's accusers. But the next paragraph of Heslop Harrison's letter to Riley reveals how personal the attacks had become. Even if they were made behind his back, they must have reached him somehow and can only have been extremely hurtful:

The 'encirclement' yarn . . . does not explain all the curious chain of discrepancies to which I have drawn attention. Everyone would laugh here (no, I mean at Newcastle!) at this quaint notion. I am happy to say I am good friends with the whole of my 150 or so colleagues and I will guarantee that not one would connect me and my psychology with Hitler's inanities. That sort of a trick phrase may be fashionable, but it is only a phrase, explaining nothing, meaning nothing and least of all warranting the perpetuation of an injustice on one, who on your own showing has given quite a lot of specimens and expected (and received!) nothing in return. But perhaps the 'lost' – no, I had better not write the bitter phrase which expresses what I feel.

I tell you frankly that I know why the tale is kept going, for I have the exact names of persons who have given it recent momentum, and I tell you just as assuredly it will neither influence me nor any of my colleagues to diverge a hair's breadth from the course we have mapped out.

I have written in full not in any way intending to be unfriendly but because I feel that the collection of phrases with which you have presented me are utterly unworthy and are no explanation of what has happened.

In spite of all this you will find me just as loyal and helpful a colleague as before . . .

Yours sincerely,

J.W. Heslop Harrison

In the same spirit that was to motivate Raven in his correspond-ence with Heslop Harrison nine years later, and even using the same phrase, Riley now realised that he had done all he could to make his points. On 31 May 1939, he wrote:

My dear Harrison,

I feel very much like the editor who writes 'this correspond-ence must now cease', for I feel that really no useful purpose will be achieved by continuing along the present lines. All the state-ments I made in my last letter were most carefully checked against the original documents here. I shall be very happy indeed to show you these next time you are in town.

I wish you success in this year's expedition.

Believe me,

Yours sincerely,

N.D. Riley

The correspondence didn't quite cease – there was one more outburst from Heslop Harrison five months later, worrying away – again – at the 'lost, broken, or never collected' rumours, but the whole file made clear that at least ten years before Raven set out for Rum, Heslop Harrison was very aware that senior colleagues in his own field were deeply suspicious of his findings

It seems to me that this correspondence highlights two factors, which, if each had existed alone in Heslop Harrison, might never have led to the state of affairs that finally prevailed. First, there was Heslop Harrison's slipshod practice as far as keeping specimens and records was concerned; second, there was his talent for making enemies, linked to a sensitivity to criticism or to the challenging of his ideas. Let us assume, for the sake of argument, that many – perhaps most – of his discoveries were genuine, but that in the mid-thirties a few were fakes in some sense; that is, they were plants or animals that had not truly grown or lived where Heslop Harrison said they were found.

If he had been meticulous at keeping his records and specimens, whatever suspect plants or animals there were could easily have been separated out for consideration, and the rest acknowledged as important discoveries deserving to stay in the reference books. *If* he had had more humility about the practice of science, about the fallibility of theorising and about the value of cautious statements as against 'know all' pronouncements, he would have had far fewer enemies interested in pursuing him, questioning his data, and spreading rumours about fakes and fraud. It may even be that the dubious records would have been quietly ignored before they became *causes célèbres*, and that Heslop Harrison himself – had he been a different personality – would have taken Riley's advice, which I think is capable of being read not as a warning about how things could *seem* to others, but as a warning that he had been found out and should stop his dubious practices there and then.

As it was, the combination of poor practice and an irascible personality was combustible, and only needed a whiff of fraud for the whole situation to burst into flames.

John Morton accepts some, but not all, of this characterisation, as he told me: 'Any suspicion of this sort spreads rapidly through the rumour mill and is almost impossible to stop. The Prof had many enemies because he was so dogmatic and outspoken and devastating in his criticism of other people. In particular he had earned the enmity of some people in his own department at Newcastle. He ruled this department as a despot. Also he had crossed swords with some of the most powerful people in British botany. Earlier in his career he had bitter battles in the Royal Society over Darwinism and Lamarckian theories of inheritance and evolution, hence there was a fertile field for rumours to spread and accusations to grow. Do I believe that he could or would fake discoveries? No, frankly I don't. Furthermore there was no need to. He discovered so many things, plants and insects; most have not been questioned. The article by Raven was damning and libellous.

Once accusations of this sort are made, it's impossible to erase the suspicions that ensue. As for *Carex bicolor* being planted on Rum, nothing planted was likely to survive on that bleak ridge where the Prof said it occurred. Faking records or results would be a heinous sin in the eyes of any scientist.'

Those are the words of a friend, the man who also told me that 'the Prof was admired by many but loved by few'.

This seems to me an extremely charitable assessment, and just how little Heslop Harrison was even admired by senior figures in the botanical establishment became clear some years after the first edition of this book was published, when a file of previously un-archived papers came to light in the Natural History Museum.

Epilogue

Two Minor Mysteries

When the first edition of this book was published, reactions among botanists fell into two main groups. There were those who said, 'We always knew it,' and others who said, 'This is a slur on a great man.'

A review in *Natural Science* seemed to say both at the same time: 'Questions about Heslop-Harrison's work in Rum signify no great scandal. Scholars were not seriously misled, with the possible exception of Heslop-Harrison himself, for it has been suggested that students accompanying him on field trips to Rum may have had some fun at his expense by secretly importing exotic species to the island.'

Some were outraged that I, a non-botanist, had even dared to write the book. Said one: 'Annual plants can pop up in any possible or impossible places and their occurrence or absence must be evaluated with great care and good understanding of their ecology. I am not sure that John Raven had enough experience to do this evaluation, but I know that Karl Sabbagh does not.'

Another wrote, of Heslop Harrison: 'His powers of observation were probably greater than the vast majority of naturalists of the time. Knowing him – no that is the wrong expression – *observing* him as I did over a number of years, my personal bets would be on him finding the disputed plants where he said he did.'

Then, a few years after the first edition of this book was published, a strange coincidence occurred which brought to light yet more evidence of professional concerns about Heslop Harrison's work, and which also demonstrated a failure of nerve, or perhaps just a desire for a quiet life, on the part of the botanical community.

A friend of mine, Stephen Bann, Emeritus Professor of History of Art at the University of Bristol, happened to be on the same committee as Dr Stephen Blackmore, at the time director of the Royal Botanical Gardens in Edinburgh, and they shared a taxi from the station to the venue of a committee meeting. The topic of *A Rum Affair* came up, perhaps because that was the only writing about botany my friend had read recently – or ever – and Dr Blackmore said he knew of the existence of a further folder of letters and memos about the case which I had never seen. The Heslop Harrison story was considered so sensitive, even in 2005 when this encounter took place, (nearly forty years after Heslop Harrison's death) that Blackmore insisted that his exact description of the events that led to these papers suddenly being available should be included in any further account of the case. Here is his statement:

When Karl Sabbagh asked for information on this case at the Natural History Museum he unfortunately asked the wrong person, who in all innocence said there was none. Had he asked me as the then Keeper of Botany he would have prompted me to open an envelope I had never examined but knew was in one of several filing cabinets in the Keeper's office. After reading the book, curiosity prompted me to look and I found a significant collection of evidence on the case, which I transferred to the publically accessible Archives at the Museum. The eventual remorse of the perpetrator of the hoax is important but can never excuse what happened. Scientists do not tolerate such behaviour which, fortunately, is not heritable and passed neither to his son, nor his grandson, both outstanding botanists in their own right.'

Stephen Blackmore, Regius Keeper, RBGE.

Of course, when this news reached me, although I had long ago left *A Rum Affair* behind me, I knew I had to see the file.[65]

In it was a letter from Dr George Taylor, who was Keeper of Botany at the Natural History Museum, writing in 1981 to another botanist, describing a process he had initiated:

I prevailed on my very old friend R.B. Cooke who put down in detail what he had told me over the years and he did write rather graphic descriptions of how things had been 'planted'. He did not want anything done with this until he had died which is some five or six years ago but I deposited it where I felt it ought to be in the British Museum and should be accessible to <u>bona fides</u> students who require to know the facts in pursuit of their scientific investigations. . . . The Hebridean records and, of course, the other spurious H-H's should now surely be completely exposed. I suppose it is a question of the sins of the father being visited on the son and indeed grandson that has inhibited complete disclosure but science is above personal matters and the truth ought to prevail.

The report by R.B. Cooke is a sober litany, headed 'Confidential for the present', which gives a plant-by-plant account of many of Heslop Harrison's 'discoveries' by a botanist who accompanied Heslop Harrison on many of his expeditions and clearly became increasingly disillusioned about the validity of the claims.

- August 1936: I was shown a dozen or so scattered plants grow-ing in a small piece of disturbed stony peaty ground . . . I

could not find a single plant of the *Juncus* . . . apart from this
piece of ground despite many hours' search over ground in the
vicinity which looked similar . . .

- Rum: Since 1938 and up to 1946 when I was last on the island,
  I have failed to find *Juncus capitatus* except in 1943, when on
  the first full day of our visit I saw a dozen or more plants
  which in my opinions had been recently planted; there were
  to be seen marks which suggested a stone having been used to
  press in the soil round the roots

- Prof. Harrison showed some of us a very withered specimen
  of *Cicendia pusilla* and said 'see what I have found'. I never saw
  a fresh specimen and was never able to find out just where on
  the shore of Loch Bee, if it was the locality, it was said to have
  been found.

- This specimen was reported as having been found among a
  number of specimens of *T. pegiferium*(?) . . . Four of us spent a
  long time searching the supposed area but saw no signs . . . it
  would seem better to ignore it.

The list goes on and on, of plants which no one but Heslop
Harrison ever found again. In some cases, plants which might be
seen as weeds in other parts of the country were only unusual when
discovered in very different soil on a Hebridean island. Cooke
wrote:

- He [Heslop Harrison] brought with him about half a dozen
  small plants or seedlings, of this *Sisyrinchium*, which he said he
  had found on Coll . . . I planted in my garden in Corbridge . . .
  the specimens given to me on Tiree by Prof. Harrison. They
  have grown well . . . I can see no difference between these
  Coll plants and those I had already in the garden under the
  same name. I have not been back to Coll but I have been told
  that Prof. Harrison says the *Sisyrinchium* is now extinct there.

There are two sad notes in the correspondence in this file. One letter suggests that towards the end of his life Heslop Harrison tried to make amends in a small way for his skulduggery. He had been sent the proof of an article he had written about his Hebridean discoveries and according to his son-in-law he 'crossed out the false ones and substituted others which were correct. . . . You will say why did H.H. make the changes? I think because he expected the Hebridean plant records to be examined by a committee and I know he had the wind up over this.'

The other sad note is a comment by R.W. Cooke, who over the years had been a good friend and trusty colleague of Heslop Harrison in his botany expeditions. Writing to George Taylor in 1955, he said, 'You may say what fools Harrison must have thought us to be, or rather I should say I was. It was a very sad chapter in my life to be so taken in by him . . .'

In the letter containing his quotation about the file in his desk drawer, Blackmore also wrote:

My reasons for insisting that the quotation is printed in full are that otherwise the impression is created that the N[atural] H[istory] M[useum] concealed the file. It may have done so before my time when direct contemporaries of Heslop-Harrison were there (George Taylor was also a former Keeper of Botany at the NHM) but certainly did not intentionally do later on. If Karl Sabbagh, who I have never met, had asked me I would have searched out and given him the file. My other reason is more personal. Jack Heslop-Harrison, the former Director of Kew, was a good friend of mine and his widow still is. Pat Heslop-Harrison, the grandson, is also a respected botanist. However, I know that the family were always troubled by the hoax and I think some subtlety is owed to them given that they had no part in the matter. I hope you will respect these wishes.

Blackmore writes that 'the family were always troubled by the hoax' whereas as my book showed – and Blackmore says he read it – Heslop Harrison's grandson told me he knew nothing about it, but if a line can be drawn under the whole sorry tale it seems to me the Blackmore has drawn it.

I almost didn't visit Birtley. After all, what can you really tell about a long-dead resident from looking at his house and garden, particularly after they've seen two changes of tenant? But in August 1998, I had an interview to carry out in Scotland, so I drove there via Heslop Harrison's home village. Apart from the pleasure of seeing Anthony Gormley's giant 'Angel of the North' sculpture, which overlooks the village from a mound by the nearby road, there turned out to be an unexpected benefit in the form of an additional small wrinkle in the mystery of the Rum affair.

Birtley, where Heslop Harrison's father – then plain Harrison – lived, is a pleasant village whose prosperity is founded on a brickworks, an ammunition factory, and an ironworks. It was also the site of England's first venereal disease clinic. The village straddles the road to Newcastle, and many of its buildings have an Edwardian redbrick cast to them, now overlaid with the facades of tandoori restaurants and video rental shops. The main clue – a pretty obvious one, you might think – that I had to the location of Heslop Harrison's house was its address: Gavarnie, The Avenue, Birtley.

It is a rather grand-sounding address, and The Avenue turned out to be a rather grand road, at least compared with other Birtley streets. Apparently, according to a local history book, it once had gates at either end that were closed every evening to establish its status as a private road. Of all the old streets in Birtley, this is the sort of place that you would expect Heslop Harrison – if he had any social pretensions, as I suspect he had – to choose.

I walked up The Avenue, which rises quite steeply from the main road, looking at every house for the name Gavarnie, or at least for a house that might be worthy of Heslop Harrison. On neither count was there a candidate. Most of the houses were semidetached or terraced. Nice middle-class residences, probably with three bedrooms, but without the grandeur of scale I expected. And many had names: Inglewood, Fernlea, Tyneholme, and – rather unfortunately – Lockerbie, but none was called Gavarnie. The only house that fitted the image I had was a large one, but that was called Rockville and its name was permanent, engraved in the solid stone from which it was built.

I got to the top of the hill and paused. In the middle of a small patch of grass was a statue in white marble of a short, upright man with a long beard. I thought at first it was Charles Darwin – rather enlightened of the city fathers, I thought – but it turned out to be a statue of Colonel Edward Mosely Perkins, who was connected in some way with Birtley Ironworks. (I discovered later that Pevsner, in his *Survey of England*, calls this 'the funniest monument in County Durham'.) I started down the hill again, still searching, in case I had missed some clue on the way up. But there was no Gavarnie and no house worthy of the professor.

When I met Heather Marshall, a librarian at the Birtley Branch Library, all became clear. She told me that Heslop Harrison had never lived in The Avenue, even though he gave that as his address. His house, certainly a substantial one, was in Ruskin Road, a street off The Avenue, and she arranged for me to see the house and garden. Now I was still puzzled. With one exception, all Heslop Harrison's letters that I had seen had 'Gavarnie, The Avenue, Birtley' written at the top. But by rights, to be accurate, he should have put 'Gavarnie, Ruskin Road, Birtley.' There was one obvious explanation, but I could hardly believe it. The Avenue was the most prestigious street in Birtley; Ruskin Road was not. Did he change his address to impress people? Did he really say to the post office,

'Look, I know I live in Ruskin Road, but you'll get a lot of letters to me addressed The Avenue. Just pop them through the letterbox, will you?'

And yet, of course, it was entirely consistent with what we know of Heslop Harrison's character. He had done the same sort of thing with his name. In legal documents – including the deeds to the land on which Gavarnie was built – he is named firmly as J.W.H. Harrison. But over the years, he spoke and wrote his surname as Heslop Harrison and probably didn't complain when some went further and printed it as Heslop-Harrison, reinforcing the impression of an English patrician family name. At any rate that's how people think and talk of him today, with as much reason – and for the same reason – that his address was given as The Avenue. The one exception to this address policy in the Raven file was Heslop Harrison's last angry letter, which was sent from 'Gavarnie, *Ruskin Road*, Birtley, Co. Durham.' Perhaps by this stage, if he thought the game was up, there was no longer any need to stand on his dignity.

Gavarnie itself was built some time after 1928, when Heslop Harrison bought the plot of land. According to a neighbour, it was called Gavarnie after a French village high in the Pyrenees, a famous site of botanical interest, where the Heslop Harrisons spent their honeymoon – probably a good place to hunt for rare Arctic-alpine plants.

The plot of land was used to build two houses: Gavarnie, for the Heslop Harrisons, and a bungalow, for his brother. The house is now lived in by a local doctor and his family, and as his wife showed me around the house and garden, she was full of praise for the way it was designed. 'It's brilliant,' she said, 'lots of little windows just where you need them and very big landings. And there's an attic where he kept his butterfly collection. The man over the road said that he used to come and visit as a boy, and the professor showed him his butterflies.'

The garden was surprisingly small. After reading about how much of Heslop Harrison's botanical and entomological work was carried out at home – including Valentine's phrase, 'Harrison has a large garden at his private house in Birtley' – I was expecting some large, rambling area with space for beds of different plants and maybe a greenhouse or two.

Where the professor grew botanical curiosities (and rare Arctic-alpine plants?) there is now a small play area for the doctor's sons. But in one corner, overgrown and not the focus of much horticultural attention these days, is a rock garden that – who knows? – still contains plants from Heslop Harrison's time.

From the industrial villages of County Durham, I headed north to Scotland. It was the best sort of weather in which to see the Highlands. No monotonous grey skies (or monotonous blue skies, for that matter), but instead scudding clouds, some bright white, against a pale blue background. The road from Callander to Oban was flanked by multicoloured hills, made even more variegated by the changing patterns of sunlight. As I meandered along behind two huge trucks, knowing there was going to be no opportunity to pass for the next twenty miles, I thought about how the hills with all their variety of colour concealed more than they displayed. To a botanist, what was significant was invisible. The patches of colour were no more than fleeting combinations of plant colour, intensity of sunlight or shade, soil chemicals, and seasonal variations. What really mattered could be understood at a distance of inches rather than furlongs.

My story, too, had its patches of brightness and its areas of shade, and the significance of each could be discerned only on a much closer inspection than I had been able to make, or than anybody could now make at this distance in time. What had become clear as sunlight was that Heslop Harrison was believed by most botanists who had heard anything at all about his work to have faked his results. What was more obscure was why he did it – if he did. And buried even deeper was the question of why everybody believed he had done it if he hadn't.

When I reached the west coast of Scotland, near Oban, the trucks went south and I drove north, past views of the Isle of Mull and of Morvern, the peninsula where Ardtornish, John and Faith Raven's estate, stands. I had arranged to see Peter Wormell, who had been Warden/Naturalist of Rum from 1957 to 1973. He was probably the last person to spend time with Heslop Harrison on Rum when, at the age of eighty, the professor returned to the island alone.

Wormell is a trim, friendly man, his face ruddied by long periods out of doors. In 1957, when he was twenty-seven, he returned from forestry work in Africa and took a job on Rum. He is now retired but is still closely involved with survey and data-collection work in connection with the plants and insects of the island. When I got past his friendly dog and into the living room of his bungalow, with picture windows looking across the water, Wormell showed me a pile of papers he was working on, a comprehensive account of the insect life on Rum, gathered with the help of a team of entomologists.

The detailed descriptions of spiders, beetles, and flies were yet another illustration of the grip that observing, describing, and surveying the smallest elements of nature have on some people. The work the naturalists had carried out recently on Rum also demonstrated one factor that bedevilled attempts to describe all of Heslop Harrison's surprising discoveries as fakes: when you might have thought that the island had been thoroughly combed from north to south, new discoveries kept being made. 'I found a flea new to science on a Manx shearwater,' Wormell said proudly.

Wormell is clearly torn in his views of the Heslop Harrison affair. He liked the professor and marvelled at his natural history expertise: 'He taught me a great deal about the flora of Rum, more than anybody else at that time. He was a very good teacher, he just put things over so well and told you all about them. And then he could sit down and reel off genuine lists of facts about plants and Lepidoptera.' Over the years, Wormell met and married a girl from Rum, Jessie, who worked in the post office on the island and who knew and liked all the Newcastle team who came back regularly each year and camped in tents or cabins.

But Wormell had to admit that there were problems with Heslop Harrison's records, problems with which he was still wrestling thirty years after the man died. As the doubts grew in the 1950s about Heslop Harrison's discoveries, the reference books would put square brackets around plants that he had 'discovered' unless they were independently corroborated. Then, as the years passed without corroboration, the plants would be quietly dropped. Wormell was now facing the same situation with the Rum insects, some of which the professor had never even published, although he told other entomologists about them. With decades having passed since some of the professor's discoveries had been announced, surely it was time to drop them from the records? But Wormell was unwilling to do so: 'There's so many of his records you can't really go along with, but so much of his work was genuine and good. I don't like to leave his records out because he hadn't published them, but they are not to be accepted in the entomology of Rum as genuine records until they are confirmed, because of the lack of firm evidence. I know that quite a number of entomologists say that we should leave them out because he was challenged, but I knew the professor, and I knew that a great deal of what he produced was absolutely genuine.'

The picture is complicated by the fact that, from time to time, some Heslop Harrison discovery is confirmed. 'He mentioned, for instance, there's a moth called the Yellow Ringed Carpet, and he got it as a caterpillar, he said, on mossy saxifrage, but Ted

Pelham-Clinton said that it usually grows on the succulent leaves of the purple saxifrage and seems unlikely to occur on mossy saxifrage. I went up searching for mossy saxifrage, got a caterpillar, bred it out, and it was exactly in the locality that Heslop Harrison said. It wasn't then considered to be a very common moth.'

Then Wormell told me a story that seemed a fitting end to my quest, a story that, in a sense, gave Heslop Harrison the last laugh. 'August 1961 was his last visit to Rum. He'd been staying with Duncan McNaughton, who had been head stalker on the island since 1913 and was still living there in retirement. The McNaughtons were going away on holiday and the professor wanted to stay a few more days on the island, so Jessie and I said we'd be happy for him to stay with us. So I took advantage of his knowledge of the island to get to know the locality of some of the interesting plants. Now, he wasn't really fit to go right up to the Arctic-alpine flora, but he did express a wish to visit once more the corrie above Kinloch where he was supposed to have found the *Carex bicolor*. As it happened, I was expecting a team of entomologists for the day, and I'd been asked to take a lot of grass cuttings up to Hallival to set as beetle traps for one of the coleopterists. So I said, "I've got to go to Hallival and I'll take you up there so long as you don't move from the corrie until I come down. I won't be very long." I just left him there to potter about, and then I went on up the hill. Then I became worried because I saw the clouds coming on over Hallival, pouring down, filling the corrie, and the mist was in no time down below and I thought, "Oh, my, where's the Prof?" So I set my beetle traps and headed back down to Coire Dubh. I had a couple of people with me, and we went up and down the corrie and couldn't see him anywhere, so there was nothing to do but carry on down and raise a search party. He wasn't very steady on his feet at that time, but, in fact, by the time we got down he'd cut across the hills and gone straight down towards the White House. I found him sitting by a big roaring fire fast asleep in an armchair and in his hand was a polythene bag with two plants of a sedge I'd never seen before. It was *Carex bicolor*, one of which had seed heads on it. He'd been on the island a week before that and when he woke up he said that there were still quite a number of plants on the site and explained exactly where he'd got them. We went back there but didn't find it, and it's never been found since.'

And perhaps that's where we should leave the professor – accomplished fraud or distinguished scientist, probably both – sitting, ambiguous as ever, in front of a roaring fire in the White House on Rum.

Except that, in this story that seems to keep providing revelations, I came across one fitting conclusion to this book in February, 2016, when in the papers I was sent by Hugh Raven, John Raven's son, there was an intriguing letter to John from John Carter, a well-known bibliophile and fraudbuster. He was the 'J.W. Carter of the year 1924' mentioned in John's obituary.

Raven had obviously sent a dossier of Heslop Harrison stuff to Carter, who had read classics at King's. The letter was addressed to 'Dear Raven' so the two probably did not know each other, although Raven would have known of Carter's investigation of several notorious literary fakes, recounted in a book he wrote, *An Enquiry into the Nature of Certain Nineteenth Century Pamphlets.*

'Of course you ought to publish more of this fascinating story – and preferably for a wider than purely botanical public,' Carter wrote. 'Even I, who don't know one flower from another, read every word of the dossier with enjoyment.' The correspondence took place in December, 1956, the year in which Raven's book, *Mountain Flowers*, was published, so it is possible that his thoughts turned to doing some more writing, of the story which he had brought to a professional conclusion six years before. 'I know how one lacks energy to take up something long after the excitement of the chase has subsided,' Carter went on, 'But it is much too good a story to lie in your drawer.'

Six years later, I went up as an undergraduate to King's, thanks partly to John Raven who was helping to decide on admissions to King's that year. He hadn't written the story up for a 'wider than botanical public' and never did. Fifty years later, that was my task, and I hope Raven and Carter would feel that I have done justice to 'this fascinating story'.

Appendix

This is the full text of the letter Professor J.W. Heslop Harrison sent to John Raven, closing the post-Rum correspondence. It is undated but was probably sent early in September 1948:

Private and Confidential
Dear Sir,
For the past three weeks I have been suffering from severe heart trouble and am quite unfit for anything. I therefore write this final letter to state plainly what I think about your various actions.

In the first place, I consider your three letters to me to verge upon impertinence. Except for a vague reference to some individual or individuals desirous of sabotaging our work, you have supplied no indications of their meaning. If they are intended to be crude attempts to throw doubt on the status of *Polycarpon* and *Carex bicolor* on Rhum, then you are doomed to disappointment. No one ever suggested that the former plant was a true native except yourself. In fact, our original record concerned one plant only! As for *Carex bicolor* that group of plants has been under observation (except for one year) since 1943, when living plants from it were sent to Kew – of which fact a printed record exists. Since then it has been shown to a number of botanists of repute. However, save to supply certain botanists with small specimens, and to examine the roots for sedge galls, the colony has never been interfered with. The other colonies, difficult to find on the broken slopes to the southeast, have been visited less frequently.

Again, your suggestion that my article on Rhum plants was written to allay your 'serious misgivings' is highly offensive. It was written to record two plants new to the Rhum list and other facts and to oblige someone who asked for copy. Here I should emphasise that I have very strong misgivings indeed in respect to the *Juncus capitatus* stated to have occurred on Barkeval, and to the alleged *Cicendia pusilla* – which became a *Wahlenbergia*. Unless the former originated in a casual seed blown from the main colony a mile away, the inherent improbability of the occurrence makes me very doubtful. As for the second plant, the facility with which such an obscure seedling was named *Cicendia pusilla* seems very significant. In addition, your reluctance to explain why, contrary to your assurance to Lady Bullough, you took these two plants away, when, on your own showing, you considered them new to Rhum, seems to call for explanation, as also does your lack of candour in not informing me, as courtesy demanded, about these two so that I could satisfy myself. Of all the botanists to whom I have shown the treasures of Rhum, you are the only one who has not instantly, and frankly, told me of any discoveries he has made; again that seems very significant.

The lack of candour you displayed in repect [*sic*] to these two plants impressed upon me at once that a similar lack existed in connection with all your dealings with me. Let me recount these.

When three years or so ago you wrote for my [illegible] you failed to acknowledge their receipt. I was compelled to write to ask whether you had received them.

Just after the war, when you first thought of coming to Rhum, instead of asking my help straightforwardly, you approached me by devious[?] methods via one of my staff. Why?

When you wished to get 'sailing directions' for the party which was organised to visit Rhum in July why did you not approach me directly yourself?

Very early a friend, who recognised the breach of professional etiquette in the organisation of such a party to work ground over which we had spent so much research effort, gave me some facts about the personnel. Why did you or its organiser inform me about its composition?* The failure to supply such information was the reason for my stipulation that any directions I gave were for your sole use, and that no information, directly or indirectly, was to be given to anyone else.

The list of personnel given later to me in your letter of May 3rd neither agrees with that supplied by my friend nor with the details you gave in your letter to Lady Bullough, when you asked for permission to land; why? The latter two <u>are</u> in agreement.

After I had gone to all the trouble to arrange with Duncan McNaughton about the dates, etc., of your visit, in your note to Lady Bullough, irrespective of any resulting inconvenience to us, you suggested that you might come on July 26th. Why?

You managed to convey to Lady Bullough the impression that I was responsible for your appearance on the island. This caused Mrs Harrison and myself extremely great embarrassment as one person on the island (not Lady Bullough or any of my party) had an objection to you personally. In spite of that both of us showed you the greatest kindness and consideration.

Ignoring any arrangements I had made on your behalf, or any possible effects of your actions on my people, in your letter of July 25th from Harris you proposed to bring two other people with you; was that a proper proceeding?

Although I showed you the plant localities you proposed to use this information in a research report; is such a thing done?

---

* Presumably, 'not' has been omitted from this sentence.

You also proposed to publish about Rhum plants. Again is such a thing done? And is it in accord with your strict undertaking not to convey information gleaned from me, directly or indirectly, to any other person?

Even your letters of August 13th and Sept. 5th show discrepancies.

In order to prevent the finding of any more mare's nests let me tell you that the *Salix* [illegible] colony was, in the end, discovered, and specimens sent to Sledge; it had been carried downhill and partially overwhelmed by a landslide. Further, I know of the existence of a plant of *Astilbe thunbergi* in the Kinloch Burn just below the *Polycarpon* rock as well as of odd plants of Herb Robert and Garlic on the Barkeval slopes. Moreover, the Kinloch colony of *Saxifragei aizoides*, which it is just possible you struck, has growing amongst its members *Heracleum sphoidylium* and *Poa annua*!

Finally, as a result of my realisation of all these facts I have come to the conclusion that you came to Rhum prepared to find fault, and therefore, quite naturally, but without the slightest justification, managed to do so.

Yours sincerely,

J.W. Heslop Harrison

Notes

1. 'John Heslop Harrison', *Biographical Memoirs of Fellows of the Royal Society*, vol. 14 (November 1968): 243.
2. John Raven and Max Walters, *Mountain Flowers* (London: Collins, 1956).
3. *Journal of Botany*, vol. LXXIX (July 1941): 112.
4. David Elliston Allen, *The Botanists* (London: St. Paul's Bibliographies, 1986): 63.
5. Charles E. Raven, *John Ray, Naturalist* (Cambridge: Cambridge University Press, 1942): 68.
6. Ibid., 81.
7. Ibid., 81–3.
8. J. W. Heslop Harrison, 'The Passing of the Ice Age and Its Effect upon the Plant and Animal Life of the Scottish Western Isles,' *The New Naturalist* (London: New Naturalist Publications, 1951): 83–90.
9. C. D. Pigott and S. M. Walters, 'On the Interpretation of the Discontinuous Distributions Shown by Certain British Species of Open Habitats,' *Journal of Ecology*, vol. 42, no. 1 (January 1954): 95–116.
10. *Vasculum*, vol. XXIV, 126.
11. *Vasculum*, vol. XXII, 62.
12. Ibid., 95.

13. All quotes from John Morton are taken from a recording he made for me of his memories of J. W. Heslop Harrison in September 1998.

14. 'John Heslop Harrison', 248.

15. The French entomologist Jean Henri Fabre – born December 22, 1823; died October 11, 1915 – provided the first detailed study of insect behaviour from his observations of insects in their natural habitats.

16. Bertrand Russell, *The Scientific Outlook* (1931), quoted in Briggs and Walters, *Plant Variation and Evolution*, 3d ed. (Cambridge: Cambridge University Press, 1977): xv.

17. *Watsonia*, vol. 15 (1984): 157.

18. The file in the Natural History Museum Library is numbered DF442.

19. W. A. Clark, B.Sc., Ph.D., 'Remarks on Certain Outer Hebridean Plants,' *Vasculum*, vol. XXV, no. 3 (August 1939): 73–5.

20. See Chapter 12.

21. Clark, 'Remarks,' 97.

22. William T. Stearn, *The Natural History Museum at South Kensington* (London: Heinemann, 1981): 302.

23. R. Fitter, A. Fitter, and A. Farrer, *Grasses, Sedges, Rushes and Ferns* (London: HarperCollins, 1984).

24. In *Flora of the Outer Hebrides*, ed. R. J. Pankhurst and J. M. Mullin (London: HMSO, 1991): 53.

25. Mysteriously, although they were so thorough in points of detail in other ways, both Maybud Campbell and John Raven spell Heslop Harrison's son-in-law's surname with an *e*, even though in all the printed papers, which they must have read, it is spelled without.

26. The abbreviation 'v.c. 110' refers to vice county no. 110, the subdivision used by botanists to refer to the Outer Hebrides. In a system devised by H. C. Watson, the counties of the British Isles were subdivided into smaller units, called vice counties, to make plant distribution analysis more precise.

27. Magnusson, *Rum*, 39.

28. Magnus · Magnusson, *Rum: Nature's Island* (Edinburgh: Luath Press, 1997): 40.

29. Ibid., 42.

30. *Sedges of the British Isles*, revised by A. C. Jermy, A. O. Chater, and R. W. David, BSBI Handbook no. 1 (Botanical Society of the British Isles, 1982).

31. Ibid., 210.

32. Ibid., 7.

33. *John Raven by His Friends*, privately printed, 1981.

34. *Journal of Botany* (July 1941): 112.

35. *John Raven by His Friends*, 89.

36. Dick David, quoted in *John Raven by His Friends*, 73: 'One does not expect a pacifist and an opponent of blood sports to be a fisherman but John, who had only fished casually and with distaste before his marriage, became an addict.'

37. When Raven wrote his first letter to Heslop Harrison, he still thought the small plant he hadn't recognised growing out of the *Polycarpon* was *Cicendia pusilla*.

38. A leading botanical journal, named after H. C. Watson, a distinguished botanist and natural historian.

39. *Watsonia*, vol. I, part VI (1950): 394.

40. Felix Franks, *Polywater* (Cambridge, Mass.: MIT Press, 1982): 172–3.

41. *Science*, vol. 244, no. 4902 (21 April 1989): 277(3).

42. *Nature*, vol. 338 (20 April 1989): 613.

43. *Science*, vol. 244, no. 4902, 277(3).

44. Ibid.

45. *Nature*, vol. 338 (27 April 1989): 694.

46. Franks, *Polywater*, 181, 173.

47. E. B. Ford, *Ecological Genetics* (London: Methuen, 1964).

48. 'Edmund Brisco Ford', *Biographical Memoirs of Fellows of the Royal Society*, vol. 41 (1995): 152.

49. J. W. Heslop Harrison, 'Melanism in the Lepidoptera,' *Entomologist's Record*, vol. 68 (15 August 1956): 172–81.

50. R. A. Fisher, 'Has Mendel's Work Been Rediscovered?' *Annals of Science*, vol. 1: 115–137.

51. *Journal of Genetics*, vol. 9, no. 3: 195–280.

52. E. B. Ford, *Ecological Genetics*.

53. This work was done away from his usual garden, which was full of other plants that would confound the experiment. So, said Heslop Harrison, 'in 1921, I secured a large garden entirely satisfying my requirements'. Later in the same article he wrote, 'I had permission to visit a very secluded piece of wasteland upon which grew a colony of *S. rubra*' (a type of willow). Clearly, if Heslop Harrison had needed to find a piece of ground in which to grow Arctic-alpine rarities, he would have had no trouble in doing so. These and other data about the sawfly work are taken from 'Experiments on the Egg-laying Instincts of the Sawfly, *Pontania Salicis Christ.*, and their Bearing on the Inheritance of Acquired Characteristics; with Some Remarks on a New Principle in Evolution', *Proceedings of the Royal Society*, B, vol. 101: 115–26.

54. *Entomologist's Record*, vol. 87 (1975): 161–6.

55. R. Colin Welch, *Proceedings of the Royal Society, Edinburgh*, vol. 83B (1983): 506.

56. G. Heslop Harrison, 'Coleoptera from the Outer Hebrides,' *Entomologist's Monthly*, vol. 72 (1936): 52.

57. Frank Balfour-Browne, *Water Beetles and Other Things* (Dumfries, Scotland: Blacklock Farries, 1962).

58. R. J. Pankhurst and J. M. Mullin, *Flora of the Outer Hebrides* (London: HMSO, 1994): 1.

59. Ibid., 2.

60. Ibid., 37.

61. Ibid., 52–3.

62. Ibid., 77.

63. Ibid., 79.

64. This can be found in the Natural History Museum Library in a box marked DF332/4.

65. In 2008, this file in the Natural History Museum archives was identified by an accession number, 2008/14.

Index

acquired characteristics, inheritance of 176, 179, 182
The Affair (Snow, C.P.) 172–3
Agabus bipustulatus 29
Allen, David 12, 53, 54, 55, 56, 152
Alps 195, 198
Altnaharra 77
Amanita muscaria 52
ammonoids 165
Amsdell Creek, New York 165
Andreaea: A. blytii 22
    A. hartmani[i] 22
Angel of the North (Gormley, A.) 234
Anglesey 77, 109
Arabis petraea 94, 95
arctic scurvy grass *see Cochlearia arctic*
Ardtornish 207, 208, 238
Arenaria norvegica (Norwegian sandwort) 22, 101, 106, 125, 126, 132
Arisaig 97, 125
Asplenium viride 10, 17
Astilbe thunbergi[i] 246
astronomy 12, 149
Aulonogyrus striatus 195
autonomic nervous system 159, 163
Aviemore 28
Ayr 191
Ayrshire 89

Backhouse, James 32
Balfour-Browne, Frank 194, 195, 198, 199, 200, 201, 202
Balkans 195
Bann, Stephen 230
Barkeval on Rum 77, 108, 115, 117, 128, 131, 133–4, 244, 246
Barra 65, 66, 83, 194, 199, 216
Bee Loch 332

beetles 42, 43, 115, 189–94
    see also specific common names
behaviorism 160
Beijing Institute of Palaeontology 169
Belted Beauty moth 201
Ben Laoigh 34
Ben Lawers (and range of) 94, 95, 112
Ben Nevis 112
Berry, Bill 166
Bethnal Green 34
Betula nana 61
biofeedback 158–160
biogeography 7
Birtley 44, 57, 88, 117, 132–3, 142–3, 234–5
Black Corrie 206
Blackmore, Dr Stephen 230, 231, 233, 234
Blake, Sexton 40
Blyth 30
Blyton, Enid 25
Bolton, Ethel 216
Botanical Society of the British Isles (BSBI) 52–6, 153, 212, 249
The Botanists (Allen, D.E.) 54, 247
botany 9–18
    classification in 9–10, 12, 14
    contribution of amateurs to 9, 12, 31–2
    gardening *versus* 11
    seventeenth-century study of 12
    during World War II 12, 43–4
Boucot, Art 169
Boyd, Morton 86
Braemar 28, 34
Brandza, Demetri 51
Brener, Jasper 160, 161, 162
Briggs, Mary 212
Brindled Beauty moth 202
Bristol University 230
British Herbarium 52, 54, 70

British Museum 208, 231
British Museum of Natural History 50, 54–5,
　　71, 126, 131, 208, 214, 217, 228, 230–31
　archives of 54–5, 126, 217, 230–31
　Department of Botany 49–50, 69n, 208, 218
　Raven's deposit of specimens in 125–7,
　　131–2
British Water Beetles (Balfour-Browne, F.) 194
Bryce (butler) 119, 120
Bullough, John 87, 88
Bullough, Lady Monica Charrington 90, 91,
　　99, 100, 101, 103, 104, 137, 149, 187
　friendly welcome for Raven from 106–7
Bullough, Sir George 87, 88, 89, 90
Bullough family 89, 90, 91
Bunge, Alexander von 51
Burton, Sir Richard F. 48
Butler, Professor J.R.M. 78, 117
butterflies 43, 175, 189, 192, 201, 218, 236

Caenlochan Glen 27
Cairngorms 27
Calgary 168
California, University at Berkeley of 166
Callander 237
Cambridge 33, 46, 75, 98, 125, 127, 130, 144,
　　149, 151, 204, 205, 207
Cambridge Botanic Garden 63, 130, 133
Cambridge House (London) 37
Cambridge University 131, 208, 247, 248
　Christ's College 12, 126
　Corpus Christi College 14
　Department of Minerology 144
　King's College 1, 4, 37, 151, 204, 242
　School of Botany 127, 130
　Trinity College 4, 8, 37, 69, 70, 76, 79, 102,
　　123, 136, 141, 144–6
Cambridge University Press 208
Campbell, Dr John Lorne 91, 92, 99, 100, 187,
　　188, 189, 203, 218
Campbell, Maybud 52, 53, 54, 55, 56, 62, 65,
　　68, 79, 85, 86, 90, 92, 97, 205, 221
Campion, W. 187
Canaries 119, 194
Canna 91, 99, 100, 101, 102, 187
Canna, Laird of 186
Canterbury Cathedral 2
Carex: C. *bicolor* 6, 9, 10, 22, 24, 68–9, 77, 146
　148, 150, 152–4, 156, 194, 209, 212,
　228, 240–1, 243
　C. *capitata* 6, 22, 71–2, 73n, 77, 116, 212
　C. *chordorrhiza* 77, 93, 96
　C. *glacialis* 22, 77, 108, 117, 212
　C. *lachenalii* 77, 93, 96
　C. *limosa* 97

C. *microglochin* 34, 77, 93, 94–6
C. *pedata* 71, 74
C. *rigida* 94
C. *rupestris* 34
C. *saxatilis* 94, 95
C. *ustulata* 94
　discovery of 108, 111, 114, 115, 119
Carlton House 39
carrion beetles, 209 192–3
Carter, Dr H. Gilbert 119, 155
Carter, John W. 3, 241
Catcheside, Dr David 119, 127, 130
cecidomyiids 143
Channel Islands 76, 77, 109, 110
Charrington, Monica née de la Pasture 89
　see also Bullough, Lady Monica Charrington
Chaucer, William 40
Chelsea Physic Gardens 50
Cheshire 142, 143
China 51, 168–9, 182
Church of England 46
Cicendia pusilla 77, 78, 118, 134–5, 138, 232
Clapham, Arthur R. 75, 79, 92, 97, 115, 205
Clark, W. A. ("Willie") 59, 60, 61, 67, 68–9,
　　86, 95–6, 110, 115, 153, 154, 155, 190,
　　198, 215, 216, 221
　and Cooke's brief against Heslop
　　Harrison 115, 153
　Davison and 59–60, 61
　and Raven's accusations 95–6, 115
　specimens donated Natural History Museum
　　by 216
　Wilmott attacked by 59–60, 66–7
Clearances *see* Highland Clearances
Clutton-Brock, Tim 11, 38, 39, 145, 156, 157
Cochlearia arctica (arctic scurvy grass) 22
Coire Buidheag, Ben Lawers 77
Coire Dubh 240
coleoptera 190, 192, 250
Coll, Isle of 66, 224, 232
conodonts 165–6
Cooke, Randall B. 68, 69, 86, 115, 153, 231,
　　232
Cooke, R.W. 233
Corbridge 232
Cornwall 77, 110, 189
Cornwall, west of 77, 109
Corpus Christi College, Cambridge 14
Cotoneaster simonsii 141
Creighton, Tom 38, 98–104, 106, 107, 111,
　　112–17, 119–24, 207
crinoids 168
Currie, Andrew 85, 86
Cystopteris montana 34

Dalmally 34
Dandy, J.E. 62
Darwin, Charles 176, 177, 179, 235
    theory of evolution of 176, 178, 180, 182,
        183
David, Dick 208, 209, 211, 212
Davidson, John 204
Davison, Alan 58, 59, 60, 61
Devon 77, 110
DiCara, Leo 159, 160, 161, 162, 163, 164, 171
Dickens, Charles 40
dicots 15
Ditrichum vaginans 22
Dobzhansky, Theodosius 175
Docwra's Manor, Shepreth 98, 207
Dorset 77
Druce, Francis 50
Drury Lane Theatre 55
Duff, Professor P.W. 128
Dundee University College 40
Durham 63
Durham, county of 235, 237
Durham, University of 3, 5, 7, 8, 39
    Philosophical Society, *Proceedings* of 200

Ecological Genetics (Ford, E.B.) 249, 250
Edelsten 221
Edinburgh 88
Edward VII 89
Edward VIII 202
Eigg 66, 83
Einich Glen 28
Eliot, Lewis 172
Entomologist (journal) 218
Entomologist's Monthly (journal) 194, 250
Entomologist's Record (journal) 187, 250
entomology 12, 39, 184, 196, 197–8, 200, 218,
        220, 239
Epilobium: E. alpinum 94, 116
    E. lactiflorum 6, 76, 94, 108, 111, 116
Erigeron uniflorus 6, 22, 76, 108, 117
evolution 176
    Darwinian theory of 15, 176, 179, 180, 182,
        227
    effect of ice ages on 21
    Lamarckian theory of 180–85, 227
Fabre, Jean-Henri 46, 248
Fiachanais Loch 101, 120
Fionchra 106
Fioray 66
Fisher, Ronald 175, 183
fleabane 22, 77
Flora of Rum (Heslop Harrison, J.W.) 92, 141
Flora of the Outer Hebrides (Pankhurst, R.J.) 65,
        86, 214, 248, 250

The Flora of Uig (Campbell, J.L.) 86
Foggitt, Gertrude 54
Ford, Edmund Brisco 175, 176, 183, 187, 249,
        250
Forster, E.M. 4
fossils 61, 165–7, 169
Foster, Garth 189, 191, 192, 193, 194, 195, 196,
        198, 199, 200, 202, 203
Fox, Reverend H.E. 31
Franks, Felix 157, 171
Fuday 66

Gavernie, Birtley 234, 236
Geller, Uri 186
genetics 39, 181, 183–5
geology 12, 22, 24, 40, 86, 149, 185
George V 202
George VI 202
Gilbert, Oliver 61
glaciation *see* ice ages
Glas Maol 28
Glen Affric 74, 75, 77
    botanical excursion to 74
Gloucestershire 189
Gormley, Anthony 234
Grantchester 14, 62
graptolites 166–7
grasses 6, 15, 16, 42, 104, 115
    flowers of 16
    see also Poa
The Great British Fern Craze (Allen, D.) 53
Greece 132
Greenland 212
Gupta, Viswa Jit 165, 166, 167, 168, 169,
        170

Haldane, Professor J.B.S. 181
Hallival 240
Hamilton, Walter 97, 99, 100, 102
Harris, Isle of 92, 93, 94, 95, 96, 97, 102,
        117
Harris Bay 101
hawkweed 28, 30, 31, 32
Hebrides 19, 21–3, 54, 56, 68–9, 70, 81, 85, 90,
        124–6, 128, 132, 151, 171, 187, 191,
        194, 196, 200, 209
    see also Inner Hebrides; Outer Hebrides
Heracleum sphoidylium 246
Herniaria ciliolata Melderis 215
Heslop, Harrison, John ('Jack,' son) 210, 212,
        213, 214
Heslop Harrison, Dorothy ('Dollie,' daughter-
        in-law) 190, 191
Heslop Harrison, George (son) 59, 190, 193–9,
        201, 202

Heslop Harrison, Helena ('Dollie,' daughter) 59, 68, 69, 86, 154
Heslop Harrison, Jack (son) 195, 198, 199, 210, 233
Heslop Harrison, J.W.H. and Blackburn, K.B. 215
Heslop Harrison, J.W.H. and Bolton, E. 216
Heslop Harrison, J.W.H. and Riley, N. 218
Heslop Harrison, Mrs 81, 207, 210
Heslop Harrison, Pat (grandson) 210, 211, 233
Heslop Harrison, Professor J.W. 3, 5, 122, 124–5, 127–32, 133–44, 217
    and access to Rum 84–6, 88, 90, 92–7, 99, 101
    botanical community, response to Raven's allegations against 77, 119–20, 153–4, 155–6, 158, 214, 230
    class background 6–8, 37
    controversial records 218
    correspondence of Raven and 102, 104–6, 108–11, 114–18, 120
    criticisms in botany journals 61–3
    David on 208–9, 211–12
    death of 187–8, 208, 214
    documents found latterly in Natural History Museum archive, Large Blue 'yarn' and 219–28
    doubts about research of, seeds of 156–8, 161–2, 164, 166–7, 169–71
    education of 8
    effect on young botanists of allegations against 57–8
    enmity of colleagues towards 46–7, 56, 85–6, 200, 221, 226–7
    fraud allegations, quietening down of 204–5, 208–10, 212, 214–16
    home of 234, 237
    inquiry, fields of 9, 10, 12, 15, 17, 18, 19, 20, 21, 22, 23, 24
    insect research by 7, 12, 19–20, 39, 43–6, 86–7, 137, 142, 175, 184, 189, 194–5, 199–200, 201, 219–28
    Isle of Harris discoveries of 92–7, 101
    minor mysteries remaining 229–33, 235–7, 239, 240
    muddleheadedness, allegations of 174–6, 178, 180–88, 190–92, 194–5, 200, 203
    obituary of 39–40, 42–3, 45–6, 184, 185–6, 208, 218
    observational skills 15
    old age 155, 208, 238
    Pankhurst on 214–16
    periglacial survival theory of 23–4
    personality of 25–6, 30, 40–41, 43, 57–8, 139, 152, 188, 227
    planning for unmasking of 68, 69, 71–9
    professional status 7
    protégés of 57–9
    Raven and, relationship between 25–7, 29, 30, 32, 36, 38–47
    Raven report on 4–5, 6–7, 32, 59, 61–2, 75–9, 92–3, 95–6, 99, 125, 127–9, 130–32, 136–7, 141–3, 174, 210
        aftermath and effects of 145–55
        during stay in Rum 102, 114–15, 122–3
    Raven's letter to Nature on 6, 149–52, 167–8, 169–70, 174, 188, 195, 206, 209
    Raven's motivation in pursuing 26, 37, 98, 226
    research style 41, 44
    Riley's correspondence with 218–20, 221–3, 224–5, 226
    rumours of wrongdoing 49, 52–60, 62, 64, 65
    scientific fraud, other alleged cases of 158–70
    sources of suspicions about 37–8, 78–9, 96, 105–6, 110–15, 118
    Vasculum and 29, 59, 62, 63, 64–5
    Wilmott and Campbell and, rivalry of 52–3, 54–6, 62, 68–9, 70–75, 85–6, 90, 126–7
    Wormell on 238–9, 240–41
    writing style of 27, 29, 30, 41
Hieracium: H. holosericeum 31
    H. praetenerum 31
    H. sparsifolium 31
Hieracium: H. holosericeum, 34
    H. praetenerum, 15, 34
    H. sparsifolium, 34
Highland Clearances 87
Himalayas 165, 166, 167, 168, 170
Hitler, Adolf 225
Hollingworth, Clive 88, 89, 90
Hughes, McKenny 183
Hutcheson, Tom 61
Hydroporus foveolatus 195

ice ages 18, 19, 20–23, 85, 199, 247
Illecebrum verticillatum 216
Inchnadamph in Sutherland 101
India 165
Industrial Revolution 90
inheritance of acquired characteristics, theory of 176, 179, 182, 227, 250
Inner Hebrides 3, 5, 17, 66, 76, 83, 84, 86, 102, 149, 190, 191, 193, 201
insects, research on see entomology
Iolande 104, 106, 109, 120, 210
Iona 33
Iraq 197
Ireland 193, 200

Jackson, Dorothy 201
Janvier, Philippe 168
Jardine, Nick 127
Jermy, A.C., Chater, A.O. and David,
    R.W. 93, 95
Jermy, Clive 208
Jessie (Rum girl who married Wormell) 239,
    240
John Innes Centre 210
Jonas, Gerald 160
Journal of Botany 24, 48, 49, 59, 62, 66, 68, 71,
    86, 115, 247, 249
Journal of Contemporary History 48
Juncus: J. capitatus 77–9, 109–13, 131, 134, 136,
    206, 232, 244
    J. triglumis 94, 95
Jurassic period 166

Katkin, Edward 160, 162
Kew Gardens 69, 215, 233
Keynes, John Maynard 4
Kilchoan 79, 92
Kildonan 215
King's College, Newcastle upon Tyne 8, 44,
    62, 76
Kinloch 103, 105, 106, 111, 240
Kinloch Burn 134, 143
Kinloch Castle 83, 87, 88, 90, 99, 101, 119, 152
Kinloch Glen 134
Kinloch House 88
Kinloch River 78, 109, 117
Kirk Fell, Ennerdale 31
Kobresia 94
Kuenstler, Peter 34

Lacaita, Charles Carmichael 71
Lamarck, Jean-Baptiste 179
    theory of evolution of 180–85, 227
Large Blue butterfly 92, 186–89, 203, 217–9,
    221
Lathyrus hirsutus 35
Lawers Hotel 34
Lay Dean 204
learning theory 160
Leeds University 104
Leicester University 75
Lemche, H. 183
Lewis, Isle of 83, 86, 216
Linnaean Society 50
Lochnagar 27
Lockerbie 235
Lofthouse, T.A. 175
London 88, 200
London, University of 200
London Library 48–9

Longfield, Cynthia 63, 64
Luzula spicata 22, 94
Lychnis alpin 6, 77, 105

McBrayne 102, 103, 130
McBride, Professor E.W. 181
McIsaac, Allan 91, 99
McLoughlin (soil analyst) 144
McNaughton, Duncan 91, 102–3, 103, 119,
    120, 240, 245
Maculinea arion 186, 187, 188, 221
Magnusson, Magnus 89
Mallaig 81, 83, 99, 129
Malta 2, 132
Manx shearwater 238
Marburg 79
Marshall, Heather 235
Marxism 180
The Masters (Snow, C.P.) 172
Maynard Smith, John 180, 182, 185
McBride, E. W. 181
Medawar, Sir Peter 181
medicinal plants 50
Melandrium 46
melanism 176, 178, 181–4, 250
Mendel, Gregor 178
Mesozoic era 167, 170
Michigan 162, 163
Microcala filiformis 30
Miller, Neal 158, 160, 161, 162, 163, 190
Miman, Zhang 169
monocots 15
    wind-pollinated 16
Morocco 164
Morton, Professor John 43, 44, 46, 57, 154,
    155, 188, 189, 213, 223, 227
Morvern 238
mosses 22
moths 176, 178–9, 182, 184, 201–3, 208,
    218
    melanism in 176, 178, 181–2, 184, 250
    see also specific common names
Mountain Flowers (Raven, J. and Walters,
    M.) 14, 27, 30–2, 241, 247
Muck 66, 83
Muldoanich 66
Mull, Isle of 238
Mullach Mor 133
Mullin, J.M. 214
National Museum of Natural History,
    Paris 168
National Museum of Scotland 107n
natural history 7, 8, 17, 27, 53, 57, 84, 148, 178,
    208, 238–9
    botany distinguished from 9

British passion for 12–13
    as hobby 31–2
natural selection 15, 176
    during ice ages 23
Nature Conservancy Council 83, 85
Nature (journal) 6, 149–52, 164, 167–9, 170,
        181, 195, 209, 218, 222, 249
Nebrioporus canariensis 193–4
New Naturalist Journal 18–9, 27, 247
    natural history book series 27, 234
New York 159, 160, 162
New York, State University of (SUNY), at
        Stony Brook 160
The New Yorker 160
Newcastle Botanical Gardens 131, 132
Newcastle upon Tyne 43, 46, 197, 215, 216,
        224, 225, 227, 234
Newcastle University 3, 5, 7, 57, 58, 60, 61, 62,
        131, 151, 153, 191
    colleagues at 174, 239
    Department of Agricultural Zoology 197
    enmity of Heslop Harrison's colleagues
        at 46, 85–6, 227
    experimental greenhouse at 132
    Foster at 191
    King's College 8, 44, 62, 76
    Morton at 44
North Evans limestone 165
Northern Naturalist Union 43
Northumberland 63
Norwegian sandwort *see Arenaria norvegica*
Norwich University 210
Noterus clavicornis 190
nunataks 22, 200

O' Connor (Dr and anonymous botanist) 57–8,
        175, 210, 214
Oban 237, 238
Occasional Notes (newsletter) 62, 152–3
Ohio State University 169
operant conditioning 159
Orchis: O. fuchsii 65, 66, 71
    O. hebridensis Wilmott 66
Oregon State University 169
ornithology 12, 56
Otinonensis, 241
Outer Hebrides 3, 5, 17, 65, 66, 76, 83–7, 93,
        201, 212–13, 214, 215
Oxbridge 8
Oxford House (London) 33, 37
Oxford University 8, 75, 187, 200
Oxyria digyna 94

Pabbay 66
palaeontology 12, 164, 165, 168–70, 181

Pankhurst, Richard 214, 216
Paris 164, 168
Paris Museum of Natural History *see* National
        Museum of Natural History, Paris
Parsons, David 187
pattern recognition 93–4
Peacock, Professor A.D. 40, 41, 42, 45, 184
Pelham-Clinton, E.C. 187, 239–40
Penicillium notatum 52
periglacial survival, theory of 23–4, 177
Perkins, Colonel Edward Mosely 235
Pevsner, Sir Nicholas 235
Philadelphia 162
Pienkowski, Jan 1
Piltdown Man 170
pinks 77
Pitlochry 34
plant physiology, research in 11, 159
Plate, River 119
Poa: P. alpina 22
    P. annua 112–3, 133–134, 136, 137, 141, 152,
        246
    P. glauca 10
pollination 16
Polycarpon tetraphyllum 77, 109, 110, 125, 132,
        141, 143, 150
Polygonum viviparum 94
Polywater (Franks, F.) 171, 249
Pryor, Mark 143
Pseudohormomyia granifex 137, 141, 142
psychology, research in 124, 159, 160, 170, 225
Pyrenees 42, 195, 236

Quaternary Ice Age 20
The Quirang 10, 17

Raasay, Isle of 17, 77, 190, 199
Ramsbottom, John 49, 69, 153
Raven, Canon Charles (father) 12, 13, 24, 32,
        33, 74, 75–80, 131, 132
Raven, Faith (wife) 4, 11, 33, 36–8, 98, 122,
        127, 208, 218, 238
Raven, Hugh 195, 204, 241
Raven, John 14, 17–18, 24, 53, 64, 69–70,
        80–82, 97, 103, 108–9, 116–17, 121,
        133–5, 138–40, 144, 157, 160, 171, 180,
        187, 189, 203–5, 207, 218, 220, 227,
        229, 236, 238, 242
    academic career of 5, 36, 76
    amateur status of 9, 12, 27, 31–2, 39, 75, 122
    application for funding by Trinity College,
        Cambridge 70
    botanical community, response to allegations
        against Heslop Harrison 77, 119–20,
        153–4, 155–6, 158, 214, 230

British Museum of Natural History, visit
to 126–7, 131
on Canna 99–102
class background 8, 37
correspondence of Heslop Harrison
and 102, 104–6, 108–11, 114–18, 120
death of 98, 208–9
description of mountain flora by 10
education 7–8
on Glen Affric expedition 74–5, 77
goals for investigation undertaken by 47
on Harris 92–6
Heslop Harrison and, relationship
between 25–7, 29, 30, 32, 36, 38–47
letter to *Nature* on Heslop Harrison 6,
149–52, 167–8, 169–70, 174, 188, 195,
206, 209
marriage 11, 208
motivation in pursuing Heslop Harrison 26,
37, 98, 226
obituary 2–3, 241
observational skills 15
personality 28–9, 30
post-war collecting trip taken by 33–7
public image 37
report on Heslop Harrison 4–5, 6–7, 32, 59,
61–2, 75–9, 92–3, 95–6, 99, 125,
127–9, 130–32, 136–7, 141–3, 174, 210
aftermath and effects of 145–55
during stay in Rum 102, 114–15, 122–3
on Rum 84–5, 87, 90–92, 93–101
sources of suspicions about Heslop
Harrison 37–8, 78–9, 96, 105–6,
110–15, 118
writing style 29–30, 104–5
Ray, John 12, 13, 14
Rhodes, Miss 106, 187, 188
Richards, Dr Paul 119, 127, 130
Riley, Norman D. 218, 219, 220, 221, 222,
223, 225, 226, 227
Rockefeller Foundation 49
Rockefeller University 159
Royal Botanical Gardens, Edinburgh 230
Royal Institution 181
Royal Society 14, 39, 43, 124, 145, 218, 227,
247, 249
motto of 181
obituary of Heslop Harrison by 39, 40, 43,
175, 184–5, 186, 218, 241
Ruinsival 101, 106, 120
Rum: Nature's Island (Magnusson, M.) 89
rushes 6, 84, 248
Ruskin Road 235, 236
Russell, Bertrand 46
Rutherford, Ernest 12

Sagina 112
S. apetala 112–13, 118, 129, 133–4, 136–7,
141, 206
S. procumbens 133
Salicornia 54
Salix herbacea 94
Sandray 66
Saussurea alpina 22
sawflies 184–5, 203
Saxifraga: S. aizoides 94, 246
S. hypnoides 10
S. oppositifolia 10
S. stelaris 94
Schoenus ferrugineus 34
Science (journal) 166, 168
Scotland 68, 74, 75, 79, 137, 142, 192, 234, 237
east coast of 34
mainland of 17, 21, 76
north of 61
northwest coast of 83
Scottish Natural Heritage 83
Scottish Naturalist (journal) 140, 146, 147
Scresort Loch 103
Seaton Sluice 30
sedges 6, 9, 15, 77, 79, 84, 93, 94, 95–7, 113,
115
difficulties in differentiation of, 102 94–5
see also Carex
Sedges of the British Isles (Jermy, A.C., Chater,
A.O. and David, R.W.) 93, 95, 97,
208, 209, 211, 212, 249
Serignan 46
Sète, south of France 42
Sheffield University 75
Shoreham 35
Silman, Brian 190
Skye 83
Sledge, Dr W.A. 104, 105, 109, 111, 116, 120,
124, 129, 130, 141
Smith, Faith Hugh 207
Smith, John Maynard 180, 182, 185
Snow, C.P. 172
South London Botanical Institute 50
South Uist 65, 66, 212, 216
Speke, John Hanning 48
Stearn, William T. 70, 71
Stellaria media 141
Stioclett Loch 96
Stony Brook 160
Stornoway 216
the Storr 10, 17
Sunderland 43
Sussex, University of 180
Sweden 169
Sweet, Walter 169

Talent, John 164, 165, 166, 167, 168, 169, 170
Taunton 56
Taylor, Dr George 62, 153, 231, 233
Teesdale 60, 61
Teeside 54
Tephrosia bistortata 45
Thackray, John 217, 218
Thistleton-Dyer, Sir William F.R.S. 127
Thlaspi 105
    Thlaspi calaminare 125, 132
Thomsen, M. 183
Tilley, Professor Cecil 131
Tiree 66, 232
Todhunter, Isaac 46
Toynbee Hall (London) 37
Trapa natans 215
Trinity College, Cambridge 102, 123, 126
    Council of 69, 70, 76, 79, 136, 141, 144–6
    Library of 3, 4
Tutin, Professor Tom G. 75, 79, 132, 208

Uisgnaval Mhor, Harris 94, 96
Uist, Shetland 101
University College, Dundee 40

Valentine, Dr D.H. 131, 132
Vasculum (journal) 29, 59, 62–5
Vatersay 66
Vicia lutea 35

Wahlenbergia nutabunda 119, 130–1, 135, 143,
    152, 194, 244
Waldridge Fell 42
Wales 98, 192
Walters, Max 12, 13, 14, 16, 23, 27, 35, 52, 53,
    62, 63, 212, 213
Warburg, E. F. 75, 79
Washington State University 168
water beetles 29, 189, 191–5, 198, 200–1, 250
Water Beetles and Other Things (Balfour-Browne,
    F.) 250

Watson, H.C. 12, 65
Watsonia (journal) 54, 153
Webster, Gary 168
Wellcome Institute for the History of
    Medicine 53
Western Isles 17, 155
whirligig beetles 195
White House 240–1
Widdy-bank 60
Wigmore Hall 55
Wilkinson, Patrick 1, 4
willow herbs 77
Wilmott, A.J. 66, 71, 90, 110, 126, 130–32, 141,
    146, 149, 217, 221
    correspondence of Heslop Harrison
        and 55–65, 69, 186, 194
    friendship of Charles Raven and 126
    and Maybud as Ferdinand and Isabella of
        botany 54, 85
    Maybud's relationship with 54–6
    planning for unmasking of Heslop
        Harrison 68–70, 72–5
    and Raven's letter to *Nature* 151–2, 167
    and Raven's trip to Rum 69, 75, 92, 102
    rumours of Heslop Harrison's wrongdo-
        ing 52, 54–6, 59, 60, 62, 65, 66–7
    territorial rivalry of Heslop Harrison
        and 54, 56, 137–8
    Vasculum paper attacking 62–3, 65–7
Wise, T.J. 3
World War II 12, 37, 43, 49, 74, 225
Wormell, Peter 238, 239, 240
Wynne-Edwards, Professor Vero 140, 141, 142,
    146, 147, 148, 174

Young, Donald 152

Zhang, Miman 169
Ziman, John 157
zoology 12, 24, 197